JOHN BERTRAM ADAMS

JOHN BERTRAM ADAMS
Engineer Extraordinary

A tribute by

Michael C. Crowley-Milling

Gordon and Breach Science Publishers
Switzerland • Australia • Belgium • France • Germany • Great Britain
India • Japan • Malaysia • Netherlands • Russia • Singapore • USA

First published 1993
Second printing 1994

Gordon and Breach Science Publishers

Y-Parc
Chemin de la Sallaz
1400 Yverdon, Switzerland

Post Office Box 90
Reading, Berkshire RG1 8JL
Great Britain

Private Bag 8
Camberwell, Victoria 3124
Australia

3-14-9, Okubo
Shinjuku-ku, Tokyo 169
Japan

12 Cour Saint-Eloi
75012 Paris
France

Emmaplein 5
1075 AW Amsterdam
Netherlands

Christburger Str. 11
10405 Berlin
Germany

820 Town Center Drive
Langhorne, Pennsylvania 19047
United States of America

Library of Congress Cataloging-in-Publication Data

Crowley-Milling, M. C.
John Adams, engineer extraordinary : a tribute / by Michael C. Crowley-Milling.
p.. cm.
Includes index.
ISBN 2-88124-875-6. -- ISBN 2-88124-876-4 (bpk.)
1. Adams, John Bertram, d. 1984. 2. Nuclear engineers--Biography.
3. Physicists--Biography.I. Title.
TK9014.A33C76 1993
621.48'092--dc20
[B] 93-42697
CIP

Contents

List of Figures vii
Foreword xi
Preface xiii
Introduction xv
1. The Beginning 1
2. Wartime Radar 7
3. Harwell 17
4. The European Element 27
5. John at CERN 37
6. CERN Politics 53
7. Nuclear Fusion 65
8. Interlude at the Ministry 79
9. Member for Research 89
10. The 300 GeV Project 97
11. The Super Proton Synchrotron 117
12. Director-General 133
13. The International Scene and the Last Years 149
14. The Man Behind the Façade 161
15. Finale 173
Acknowledgements 175
Appendix 1 The Accelerators 177
Appendix 2 Particle Physics 183
Index 193

List of Figures

Figure 1 Fairy Farm Cottage, Mottingham 1
Figure 2 John with his childhood friend, Jim Pollock 2
Figure 3 John and Maclusky outside their lodgings 8
Figure 4 John and Renie at Malvern 11
Figure 5 The bridal pair 12
Figure 6 A page from one of John's TRE notebooks 14
Figure 7 One of John's sketches for the cyclotron magnet 21
Figure 8 The Harwell 110 inch synchro-cyclotron 23
Figure 9 John with Sir Ben Lockspeiser 30
Figure 10 The magnets in the PS ring tunnel. The proton beam from the injector comes in from the right 45
Figure 11 Mervyn Hine and Kjell Johnsen at the PS controls 46
Figure 12 Wondering why it didn't work! *Left to right:* John, Geibel, Blewett, Lloyd Smith, Schmelzer, Schnell, Germaine 47
Figure 13 John with the famous vodka bottle and the photographic evidence of the PS reaching 25 GeV 50
Figure 14 John and Bakker at an early meeting of the CERN Council 56
Figure 15 Farewell to CERN. *Left to right:* Renie, Weisskopf, Eliane de Modzelewska (Secretary to the Council), John 63
Figure 16 Luncheon discussions in one of the courtyards at the Culham laboratory 72
Figure 17 The Phoenix II experiment and the team which included two Russian scientists 76
Figure 18 A cartoonist's idea of the Ministry of Technology 81
Figure 19 On his return to CERN, John poses in front of the 300 GeV design he inherited 103
Figure 20 The Committee of Council meeting in which the new design was revealed. *Left to right:* John, Gregory (Director-General), Amaldi (Council President), Hampton (Director of Administration) 110

Figure 21 The proposed site for the 300 GeV machine alongside CERN. The existing PS and ISR rings can be seen below the main road from St. Genis (*left*) to Meyrin (*right*). The Swiss/French border is shown as a series of small crosses 111

Figure 22 Margaret Thatcher at CERN. *Left to right:* Allaby, Mrs. Thatcher, John, Pickavance, Jentschke, Flowers, Shaw .. 113

Figure 23 The machine that bored the SPS tunnel 123

Figure 24 The magnets mounted in the SPS tunnel. The upturned ends of the coils that had the insulation problems can be seen at the lower left 124

Figure 25 John makes the ceremonial final connection in the magnet circuit 126

Figure 26 The author demonstrates the control system to John 127

Figure 27 Interested observers at one of the pre-run tests of the SPS equipment. *Left to right:* Wüster, John, Milman, the author, Baconnier 128

Figure 28 Setting up for the first test 129

Figure 29 The beam goes all the way round! 129

Figure 30 Willi Jentschke is happy to pass over the Directorship of CERN to van Hove and John, in the presence of Paul Levaux, President of the Council 138

Figure 31 A gathering of CERN Director-Generals on the occasion of the 25th birthday celebrations. *Left to right:* John, Jentschke, Bloch, Weisskopf, van Hove 141

Figure 32 Site of the LEP project. The lightly shaded area shows the limestone in the foothills of the Jura mountains 146

Figure 33 John and Bas Pease with Artzimovich and other Russian delegates during the conference on fusion research in Russia 149

Figure 34 Sign outside a Chinese factory. It reads "Welcome to our uncles and aunts of the European Organization for Nuclear Research who have come to visit our children's crèche." 151

Figure 35 Audience with the Chinese Premier, Teng Hsaio-ping ... 152

Figure 36 The second visit to China. Discussing the Ming tombs site 153

Figure 37 John and Renie with two of their grandchildren on holiday 164

Figure 38 One of John's "doodles" 165

APPENDIX 1

Figure 1 (a) Vertical section through a cyclotron. (b) Plan view of "dees", showing particle orbit as a series of semicircles of increasing radius 178

Figure 2 The focusing of particles in the vertical plane by a magnetic field that decreases with increasing radius. Arrows show the direction of the force on a circulating particle 179

Figure 3 An optical analogue of alternating gradient focusing 180

Figure 4 Cross-sections of magnets, showing the pole shapes needed for alternating gradient focusing, using combined function magnets 181

APPENDIX 2

Figure 1 The change in 50 years as to what were considered to be the "elementary" particles of matter. 187

Figure 2 A bubble-chamber picture and its interpretation. The incoming neutrino leaves no trace until it interacts with a proton in the liquid, when its energy is given up in generating five charged particles, two of which decay into others while still in the chamber 189

Figure 3 Big accelerators need big detectors! The Aleph detector at LEP 190

Foreword

John Adams was indeed an extraordinary engineer. He belonged to that generation which earned its spurs during the Second World War in radar and atomic energy, at TRE and Harwell, respectively. Most of us were self-taught to a considerable extent, because the projects we were engaged in under the stress of war far outstripped what we had acquired at university. But for family reasons John's formal education did not even get that far. As a student apprentice at Siemens in Woolwich he obtained a Higher National Certificate in 1939, working part-time at the local technical college. His acute understanding of the art and practice of engineering probably owed at least as much to his father's interests in practical invention, amateur though they were, and later to the constant challenge and encouragement of physicists like H. A. B. Skinner and J. D. Cockcroft.

Adams led and largely chose the teams that built the great accelerators at CERN, and in between made important contributions to the facilities for fusion research at Culham, and to the organization of applied research through the Ministry of Technology. Throughout his career he recorded his thoughts and his proposals in copious notebooks in classical scientific fashion. Reactions to job offers are cheek by jowl with fragments of accelerator design or the outlines of laboratory policy. He successfully combined the art of delegation to tried and trusted colleagues with a meticulous attention to detail, in administration and planning as much as in engineering and design. Above all, he had to understand exactly what he was doing and why, no matter how abstruse. This was probably the basis of his success.

I knew John when we were both in Harwell, and later when he was the director of Culham. But my closest contacts with him, by now a kind and constant friend, were during my period at the (then) Science Research Council, 1967–1973, when the SPS project was being brought into existence. During that time I was the British member of the CERN Council and I often enjoyed the warm hospitality of John and Renie Adams in their Geneva home. I shall never forget the brilliance of John Adams's solution to the apparently intractable problem of how and where to build the new accelerator. Rather than continuing to grapple with national rivalries, every country insisting its favourite site be adopted, John showed that it could be built better and more cheaply at CERN itself, using many of the existing facilities. It was a *tour de force* and it won the day.

Friends and colleagues will all be grateful to Michael Crowley-Milling for bringing John briefly back to life in this sharply focused tribute to his memory—briefly, because this is a book one reads quickly, entranced by the story it tells and by the brilliance of its chief character. But the book achieves more than the informal biography it set out to be. It is also a fascinating contribution to the histories of the establishments through which John passed, and of the people with whom he worked most closely.

John Adams was an extraordinary engineer, but he was also an extraordinary European. When he died in Geneva in 1984, at the age of 64, richly rewarded with civic as well as academic honours, he was everywhere regarded not only as one of the world's great engineers, but also as a major contributor to the postwar revival of European science. His laboratories are his living memorial. He will not be forgotten.

Brian Flowers

Preface

In his later years, John Adams became a close friend of mine, and confided in me many of his hopes and fears. As one of his Group Leaders during the SPS construction and later, first as SPS Division Leader and then as a member of the CERN Directorate, during his second period as Director-General, we used to lunch together almost every day that he did not have some official function. Therefore I feel that I had a special opportunity to study the man and try to find out the secret of his success. However, John was a very private man and, while he would discuss many matters that interested him and use me at times as a sounding board for ideas, he very rarely revealed his inner thoughts.

When his wife, Renie, asked me to write this biography, I had some doubts as to whether I could do justice to the subject and turn out something that could be read with interest by a wider public than just those who knew John. It is with her constant encouragement that I have been persuaded to persevere and produce this volume.

I have tried to leave out unnecessary technical details in the main text, including only those essential for the continuity of the narrative, but have included two appendices on the machines and the physics, for those readers who have not studied these subjects and want to know a little more about the background to John's activities.

Introduction

On 3 May 1976, people started arriving in the early morning in ones and twos, and later the trickle became a stream. By ten o'clock the dimly lit control room was crowded with people. Some were there out of necessity, some by right of seniority, but others entered tentatively, afraid of being turned away, but wanting to see what might happen. They had all come in the hope of seeing the realization of a dream that began more than a decade ago. The specialists were seated in front of the control desks, with their multi-coloured computer displays, touching buttons and turning knobs to check and start up the individual pieces of equipment, silently cursing the onlookers for getting in their way. Finally, the time had come when all systems were ready. An operator spoke into a microphone connected to another control room three kilometres away, saying, "We are ready for injection." Those who knew what was supposed to happen concentrated on one particular screen. The operator touched a control. A few seconds later, there was a flash on the screen. Every few seconds it was repeated again and again.

A cheer went up and the atmosphere changed suddenly from one of ill-concealed anxiety and apprehension to one of exaltation and profound relief. The noise level rose to an almost unbearable level with mutual congratulation. Those little flashes of light, just momentary intensification of the picture on a television screen, were insignificant in themselves, but their import was dramatic. They showed that a beam of protons had, on the first attempt, travelled all the way round the largest particle accelerator in the world, built in a circular tunnel nearly seven kilometres in circumference, deep under the countryside near Geneva.

All those who had been responsible for the multitudinous components of the machine were relieved that their systems had worked, but no one was more relieved than John Adams, the director of this vast project, who had rescued it from seeming political disaster and nursed it through technical problems to a successful conclusion. This was one of the crowning achievements in an outstanding career which started in the least auspicious circumstances, went through many trials and tribulations, had many successes and few failures, and which led to recognition of his exceptional abilities, not only by the scientific world, but also by governments.

It is my hope that reading about the life of this remarkable man will encourage others, who have not had the opportunity to gain formal academic recognition, to persevere in the career to which they feel driven. I have tried, not only to record John Adams's career, but also to probe deeper to find what it was in his personality that led to his success and to justify the title by showing that he was an extraordinary engineer.

Chapter 1: THE BEGINNING

An event occurred on the 24th of May 1920 that was to have repercussions far outside the hospital in Kingston, Surrey, where it took place. This was the birth of a son to Emily, wife of John Albert Adams. The couple already had a daughter Marjorie, born eight years earlier, and the son was christened John Bertram. The family were then in what might be described as "straitened circumstances". The father had had a good job working for a London couturier until the 1914-18 war, when he joined the Pioneer Corps. In action in France, he was gassed and shell-shocked, an experience from which he never fully recovered. After the war, he found difficulty in getting a job and was unemployed for long periods. Despite these difficulties, it seems that the young John was brought up by caring parents.

Fairy Farm Cottage, Mottingham.

In the reminiscences of his childhood that he wrote shortly before his death, John described his early years as being, in the main, happy ones. When he was very young, his family had moved to a 200 year-old house, called Fairy Farm Cottage in Mottingham, a village on the Kent side of Greater London, which was rented from a relation of his mother, a corn and seed merchant. The cottage was a simple one, with outside toilet, cold water, and no electricity. It still exists, and is now a listed building. John remembered how, when he was big enough, he used to turn the handle of the big mangle to squeeze the water out of the clothes after they had been washed in the boiler in a lean-to at the back of the house. There was enough ground to grow all the vegetables they needed and to keep chickens. John's mother, "a tall woman, full of life, redheaded as a girl and with a flaring temper to match", was the strong one of the family at this time, as his father, though strong physically, "just had not the spirit nor the courage to rise from the situation, despite all the encouragement my mother provided". John often had to tramp across the fields with his father to collect the "dole", the unemployment money. At other times his father worked at the corn and seed business of his landlord.

John described his relationship with his sister as being rather remote, eight years being an enormous gap between two children, and wrote "from my point of view she was more like an *au pair* girl than a fellow child". However John was not lonely because there were children of his own age in the village, one boy, Jim Pollock, becoming his closest friend. About him, John wrote:

> Jim in most ways was the exact opposite of me. Whereas I was thin

John with his childhood friend, Jim Pollock.

> and small as a child, with a large head and a small wiry body Jim, although the same age, was larger and fatter than me, slower in movement but moved with decision in whichever direction he was going. We must have looked like the two popular comedians of that time, Laurel and Hardy. We were firm friends and spent the whole of our boyhood together.

They followed closely the cricket matches and football games that took place, according to season, on fields near to the cottage, learning to climb up and balance on the high fences intended to keep them out, in order to spectate. John claimed that the skill obtained in this climbing was to be of vital use to him later in the mountains. They also became expert in finding the lost cricket balls to use in their own totally engrossing version of the game. At other times they ventured into some large woods nearby, with "trespassers will be prosecuted" signs and imagined perils for the unwary, such as man-traps and gamekeepers behind every corner, which added spice to their adventures. There were no paths through the undergrowth and one could easily get lost. John was convinced that these adventures helped to cultivate in him a good sense of direction.

If John inherited his passion for science and technology, it must have been from his father, who had a number of interests before the war. He is credited with the invention of an automatic braking system for preventing electric trains from running into one another, similar to the system devised by Westinghouse some years later. Unfortunately, he did not patent it. The electromagnets that he had wound for his prototype system were still around and these fascinated young John, who often played with them. His father was also an expert photographer and it is clear that, but for the war, John would have grown up in very different circumstances. As it was, he attended the village school and did sufficiently well that, at the age of eleven, the headmaster suggested that he should try for a scholarship at one of the local colleges. "Both my family and Jim's had ambitions for their offspring and we seemed to at least have a kind of cockney brightness and did not do too badly at exams. I was duly entered for a scholarship for Eltham College, a boarding school for the sons of missionaries which took some day boys." He managed to pass the written examination, but then it was necessary to have an interview with the headmaster. John was very nervous and tried to conceal it by holding his hands under the desk in front of him. After the interview he found he had skinned both thumbs by rubbing them against the desk. However, the interview must have been successful, as he entered Eltham College in September 1931. There is no doubt that this was one of the turning-points in his life. As he pointed out, instead of staying at the village school till he was fourteen, as most of his friends, and then starting as an errand boy or similar, wider horizons were open to him.

At Eltham, John was a good but not academically outstanding pupil, but was very interested in art and drawing. The arts master wanted him to study architecture and even offered to lend him money to pay for training for that profession. However, after matriculating in 1936, he decided that it was time for him to start to earn his living to help out his family. By then, his mother was crippled with rheumatoid arthritis, but his father had risen to the situation and taken over the running of the household. In retrospect, John thought that it was as though he had finally found a job that only he could do and no longer felt a failure. As a result, he remembered them as a devoted couple helping each other and, given the blows that they had suffered in their lives, remarkably cheerful and loving. His sister Marjorie was already working as a shop assistant and John decided to try to make his career in the electrical engineering industry, presumably as a result of his father's earlier experiments and his delight in playing with the bits of equipment left over from them, so he enrolled as a student apprentice at the Siemens Laboratories at Woolwich. Here, he worked under the guidance of J.R. Hughes on the physical and physiological aspects of the telephone, with the aim of improving the quality of transmission.

He also enrolled at the South East London Technical Institute, where Hughes was one of the instructors, to work for a Higher National Certificate in Electronics, which he obtained in 1939. As a result of this, he was accepted as a graduate of the Institute of Electrical Engineers. One of the subjects to which he paid special attention was Transmission Line Theory, obtaining an "A" grade, and this knowledge was to be put to good use sooner than he had expected. Passing this examination completed the rather limited extent of his formal academic education, making his subsequent career all the more astonishing. His own explanation as to why he had done so well without further education was to say that, if the essence of university training is to learn from capable men, then he gained this advantage in the places in which he worked in his early years. Later he was to come into contact with many eminent scientists, who recognised his ability and gave him every encouragement, but this would not have achieved much without John's incessant urge to master each subject he came across.

It was while John Adams was at Siemens that another aspect of his character was developed. He bought a bicycle and went for long rides at the weekends, many of these trips ending up at the Youth Hostel at Canterbury, which was run by a remarkable couple, called Hemmings, with whom he became great friends. They were enthusiasts for everything concerned with the countryside and encouraged John's emerging interest in nature and its infinite variety. He has a sharp eye that noticed many things that others would have missed and, under the Hemmings' tutelage, he rapidly became expert in the recognition of the rarer species. This interest in nature became a great source of relaxation in his later, more responsible years, which allowed him to think out the problems logically when taking long walks or skiing in the Jura Mountains behind Geneva.

This relatively settled life, that was so far little touched by the war, except for the night-time visits to the family Anderson shelter, was rudely interrupted in 1940 by the bombing of the Siemens factory, which left him without a job. He registered for National Service and was sent to the Telecommunications Research Establishment (TRE), then situated in some school buildings at the village of Worth Matravers, on the south coast near Swanage in Dorset, as a laboratory assistant, grade 3.

Chapter 2: WARTIME RADAR

The Telecommunications Research Establishment had been formed to do research and development of equipment known variously by the code names of Radiolocation and RDF (Range and Direction Finding). This was used initially to determine the position of enemy aircraft, by using the reflection of radio beams from them. Later on, it was renamed radar (radio detection and ranging) by the Americans, a name which it is now universally known. Sufficient has been written about the early developments of radar to make it unnecessary to go into details here, but the situation at the time John joined TRE was that the long-range "Chain Home" stations, which could detect aircraft across the Channel, with moderate accuracy in direction and some indication of height, were in operation round the East and South coasts of Britain. There were also a number of specialized equipments for detecting low-flying aircraft and helping to aim searchlights and anti-aircraft guns.

All these systems used radio transmissions in the metre wavelength range, and the directional accuracy was limited by the wavelength and the size of the aerial. The main aim of TRE at that time was to develop radar systems working in the centimetre wavelength range to obtain good accuracy with relatively small aerials, such as could be fitted to aircraft. This posed many technical problems, as suitable components were not available and had to be developed from scratch.

All through his working life, John kept a series of notebooks which were in part records of meetings and experiments carried out and in part calculations and notes of things that interested him. Unfortunately, a number of these cannot be found, but some of the early ones are still in existence and the title of the earliest shows that it was "Started on my first joining TRE at Swanage", with the first entry dated "19 Dec '40". Subsequent entries showed that he was interested in, if not actually working on, a wide variety of subjects. One of the earliest experiments he recorded was to try to use an oscillating quartz crystal as a microwave detector. This demonstrated the principle, but proved to have insufficient sensitivity. John was then involved in the development of microwave radar for night-fighters, working for Herbert Skinner, who had studied under Lord Rutherford at the Cavendish Laboratory before moving to the University of Bristol. He had joined TRE soon after the outbreak of war, where he had earlier invented the switch which enables a single aerial to be used both for transmitting and receiving.

This new work meant going to even shorter wavelengths, requiring the use of hollow "waveguides" instead of cables for the transmission of signals. The theory of transmission of power by waveguides was only then being worked out, as an extension of conventional transmission line theory and here John's earlier studies proved useful. His contemporaries remember that John had a kind of instinctive feeling of what was needed and became expert in both the design and construction of prototypes of the components needed. The TR (transmitter/receiver) box for this microwave radar had a large number of components that had to be interlinked in the smallest possible space. Douglas Wooton, who shared lodgings with John at that time, remembers that he was so good at doing sketches and drawings that he got the job of designing the layout and fitting it all together. He said "John was obviously going to have a good career. Though he had fewer qualifications, he did not appear inferior to us graduates. He was calm and collected and just got on with the work in a splendid fashion." John's notebooks are full of sketches of various pieces of equipment and show that he was also involved in the display units to give the required information to the pilot and navigator of the aircraft, another case where three dimensional visualisation of the layout was required.

John and Maclusky outside their lodgings.

In the summer evenings, John, used to go for walks along the cliffs at Swanage, often accompanied by Wooton and Gordon Maclusky, who also shared the same lodgings. Wooton related how on one occasion they returned along the beach. They came to a barbed-wire fence with a notice on it, facing the other way. It was only when they climbed over the fence and read the notice that they realised that they had been walking through a mine-field!

Field tests of the radar equipment meant that John had occasionally to fly in a Beaufighter. On one such occasion, a door in the floor of the aircraft had not been secured properly and came open during flight. John was close to the door and only the rapid action of one of his companions saved him from falling out. He was luckier than the crew of a Beaufighter from Swanage, which had been fitted with a specially enlarged nose to accommodate a radar scanner for tests. A fighter pilot mistook it for a JU-88 and shot it down, with the loss of all the crew.

Late in 1941, there occurred a chance event that was to have an enormous effect on John's subsequent life. Forres School, near Swanage, had been taken over by the Ministry of Aircraft Production and was being used to give a three months course on electronics and radar to science graduate members of the Women's Auxiliary Air Force (WAAF) before they were sent to operational posts. One of the students to come to this school was a Miss Renie Warburton, who had gained a physics degree at Liverpool University, under Professor James Chadwick, discoverer of the neutron and Nobel Prize winner. Her meeting with John is best described in her own words:

> I first met John at a dance at the Grand Hotel Swanage one Saturday night in September of 1941. He was a tall, slim, strikingly handsome fellow with lively green eyes and an unruly mop of dark curly hair. He had a large white bandage on the first finger of his right hand.
>
> We were dancing a 'Paul Jones' and when the music stopped there he was in front of me. We greeted each other and I enquired what had happened to his finger which looked rather painful. He explained that he had damaged it at work on a lathe. The music changed and we rotated in opposite directions. When the music stopped there he was again directly opposite me. It was quite uncanny but every time the music stopped there he was. Perhaps it was providence but we had clearly been introduced.
>
> Before saying good night he asked me for a date suggesting that we might see a film together. When we were fixing a time, he said that he could not manage to come along for the supporting film, but would be there at 8 p.m. for the big feature. This should have given me a clue. On our first date he was going to work overtime!
>
> We soon found out that we both loved walking and at the weekends we often explored the cliffs on the Dorset coast and Studland Downs and we saw a good deal of each other for the next few weeks. At the end of my course we had an examination and John helped me to under-

stand some of the lectures on Transmission Line theory.

Having finished the electronics course, I was posted to 60 Group headquarters at Stanmore, Middlesex. It was war time and we had no idea what the future would hold. We were both 21, enjoying life, and meeting lots of interesting people. Thinking our paths would not be likely to cross again we cheerfully said 'Goodbye'. We did not even promise to write.

However providence played its trump card. After about a month I was posted back to Forres school as a lecturer and attached to RAF Worth Matravers. I was to help with the training of radar officers.

John and Renie resumed their walks together and saw as much as possible of each other, but before the romance could go much further, there was another interruption.

German planes had been strafing the radar stations round the South coast for a considerable time during 1941, but the Swanage area had escaped comparatively lightly. No appreciable damage had done, and the only casualty was a naval officer who had been attached to the station as an observer for the Fleet Air Arm. However, in the spring of 1942, intelligence reports suggested that the Swanage installation could come under more intensive attack, and rumour had it that a force of German parachute troops was being assembled across the Channel for this purpose. The threat was taken seriously by those in command and a whole regiment of infantry was moved in to protect TRE. This was not entirely welcome, as A.P. Rowe, the TRE director, recorded: "they blocked the road approaches to our key points, they encircled us with barbed wire, they prepared to put demolition charges in our most secret equipment and, in the exercise of their lawful duties, they made our lives a misery". Meanwhile, a search was made for an alternative site, sufficiently distant from the Channel to make a successful attack less likely. This was found in Malvern, at the College from which the pupils had been evacuated at the beginning of the war. They had returned later, but the headmaster agreed that the school should be evacuated once again, this time to Harrow, and the buildings and extensive grounds were rapidly converted for occupation by TRE. The electronics school was also moved to Malvern and Renie went there at a moment's notice when she found that some secret equipment was being driven off by a civilian driver without any RAF escort. She was assigned a billet at a local vicarage.

John was moved to Malvern a little later and was successful in finding lodging in the converted stables of a school near the vicarage. By the

John and Renie at Malvern.

25th May, the whole of TRE had been moved to Malvern and a week later Swanage was heavily bombed.

John and Renie continued to meet whenever possible, going for walks or on other trips. It was on such a trip, in the late summer of 1942, in a rowing boat on the river Wye at Worcester, that John summoned up the courage to ask Renie to marry him, and she accepted with joy. They were married at Harewood, near Bolton in Lancashire, where Renie's family lived, on the 23rd January 1943 and spent a short honeymoon at an hotel in Bakewell, Derbyshire. They did not own a car at the time but made friends

The bridal pair.

with a Ministry of Agriculture official, who took them with him to the farms where he earmarked the young bulls. Renie remembered:

> He obligingly took us around with him in his van and John helped the farmers to hold down their bulls while I sat in the kitchen drinking tea and chatting to the farmer's wife. Our honeymoon photographs are consequently unique as they are largely of farms, bulls and the occasional pig. The bride, or rather the bride's legs only appeared once when I was holding a prize ram while John took its photo. "Scientists are queer birds" as the vicar's wife so truly remarked.

Back at Malvern, they found some accommodation in a large house which had belonged to a doctor, When he died, he had left it to his spinster sister, who found it far too large for her alone, so she let the former surgery and some rooms upstairs to John and Renie.

Meanwhile at TRE, John was transferred to a group led by Arthur Starr, who had made his fortune in 1934 by writing one of the standard text books on electrical engineering, entitled *Electric Circuits and Wave Filters*, and it is an interesting coincidence that John had earlier been given this book as a prize in his final year at the Technical Institute, for "the most satisfactory performance". Whether it was due to this or not, Starr now took an almost paternal interest in John. This group was working on ASV (Aircraft-to-Surface-Vessel) radar, mainly to detect submarines that surfaced at night to charge their batteries, equipment that was to play a vital part in winning the "Battle of the Atlantic". The group was also working on equipment for the microwave transmission of speech, using pulse modulation techniques. Since microwaves could be focussed into narrow beams, this gave the possibility of providing a means of radio communication that would be difficult for the enemy to intercept. The work involved the construction of transmitters and receivers and the carrying out of field tests, which culminated in successful two-way communication between mobile ground stations at Fishguard and Aberdaron, across the full width of Cardigan Bay, in the Autumn of 1943. John was given the job of taking a van with one set of equipment to Aberdaron. He had not driven a vehicle before, but he did not let this deter him, and arrived successfully, but with the top of the van a bit crushed when he had misjudged the height of a bridge. Luckily, the roads were not very busy in that area in wartime! Later on, when the war was coming to a close, the Air Ministry decided that Starr's work would be more appropriate to the communications industry than to a laboratory concerned with air defence, and Starr left TRE and

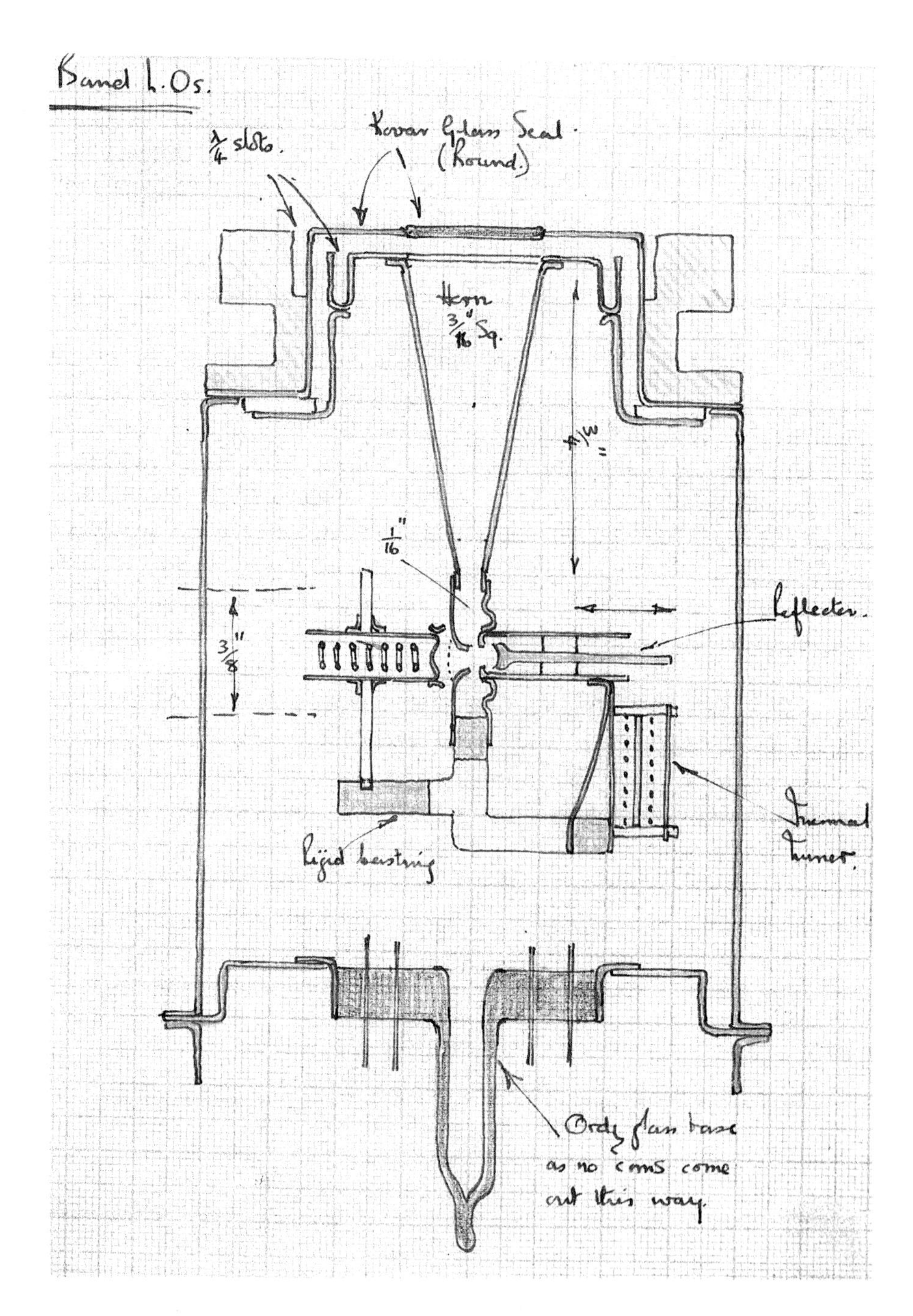

A page from one of John's TRE notebooks.

went to work at Standard Telephones and Cables (STC).

John's engineering talent and skills were soon recognised by his superiors, so that Skinner and Starr succeeded in obtaining his promotion into the Scientific Officer class, despite the lack of the paper qualifications normally required for entry. It is clear that his time at TRE was the major formative period in his life. Although he only started as a laboratory assistant, he was working with brilliant young university graduates and soon found that he could discuss technical matters with them and make useful contributions, which increased his self confidence. Mervyn Hine, who figured prominently in John's later history, was one of several young men who had interrupted their university career at Cambridge to work at TRE. Although not working on the same projects as John, he later remarked that the responsibilities that John had to take upon himself for the radar work, in his early twenties, not only in the technical aspects but also in the many varied human contacts, almost certainly determined the early maturation of his qualities as a leader. He also made many contacts that were to be useful to him in later life.

Towards the end of the war, Renie's duties in the WAAF called for her to spend long periods at RAF stations away from home, mainly for the purpose of observing the operational use of new equipment. She did not like being away from John and thought that there was not much more she could usefully do. In addition she very much wanted to have a child. She waited until she was four months pregnant and then resigned from the WAAF. Their first daughter, Josephine, was born on the 1st December 1945.

At the end of the war, many of the staff at TRE returned to their previous civilian careers and Malvern College returned from Harrow, but a slimmed-down TRE continued to occupy the barracks which had been built in the college ground and some staff had already moved to the nearby naval radar establishment, H.M.S. *Duke*. The peace-time TRE continued to work on radar improvements, but a portion of it was transferred to the Ministry of Supply to work on "spin-off" developments; a notable one being the construction of the first microwave linear accelerator (linac), using the power from a high-power magnetron, which had been designed for radar use, to accelerate electrons to high energies. I was then employed at Metropolitan-Vickers in Manchester, where I had also been working on microwave radar during the war, and became involved in this work, leading to the manufacture of linear accelerators for the treatment of cancer and for physics research. It was through this work that I became, indirectly, to be closely associated with John in later years. It is clear, from his notebooks, that John was interested in this work, although not directly involved.

About this time, John had three offers of employment. Philip Dee, a nuclear physicist who had worked with Rutherford at Cambridge before the war and had then played a very great part in the development of high-power microwave devices at TRE, had been appointed Professor of Physics at Glasgow University. He planned to build some accelerators there and offered John a post. Arthur Starr made the suggestion that John should join him at STC to continue the work on microwave communications. The third offer was from Herbert Skinner, who asked him come and carry out the engineering for the construction of a cyclotron, a type of particle accelerator, at a new laboratory that was being started up at Harwell, in Berkshire.

Renie remembers that John came home one night and said "Skinner wants me to build a cyclotron, do you know anything about them?" There had been a cyclotron in the basement at the Physics department in Liverpool but this had been out of bounds to the students. However, she had learnt enough to be able to explain that it involved building a very large electromagnet. She claims that, since these had always fascinated him from childhood, this was what decided him to accept Skinner's offer and go to Harwell. However, John was not one to take anything on without learning exactly what it involved, since one of the driving forces in his character was the urge to master any subject in which he was involved. So, when he agreed to go to Harwell to build the cyclotron, he must have found out more than just that it involved an electromagnet! He got down to work on this new proposal at once, and his notebook of that period shows that he was working on calculations concerning cyclotrons for some three months at Malvern before he and the rest of Skinner's team moved to Harwell.

Chapter 3: HARWELL

During the war, many of the most prominent British scientists, as well as some of those of other nationalities who had managed to escape from the European continent, had gone to the U.S.A. to make major contributions to the Manhattan Project, the code name for the development of the atomic bomb, and to other important activities in that country and in Canada. When the war was over, many of the British came back, but the American MacMahon Act of 1946 meant that there could be no free exchange of information about the military or peaceful use of atomic power between the two countries. Therefore the British government decided to build up its own capability in this field and, in the House of Commons on the 29th October 1945, the Prime Minister, Clement Attlee, announced that the Government had decided to "set up a research and experimental establishment covering all aspects of the use of atomic energy" under the responsibility of the Ministry of Supply.

The seeds of this enterprise had been sown at a meeting in Washington, D.C. in October 1944. This was organised by J.D. Cockcroft (later Sir John), famous for producing the first artificial disintegration of the nucleus, an achievement for which he was to receive the Nobel Prize in 1951. He had also played a large part in the production of radar equipment as Chief Superintendent of the Air Defence Research and Development Establishment (ADRDE), but was then Director of the Anglo-Canadian project on Atomic Energy, at Chalk River, near Montreal. Amongst others present were James Chadwick, Renie's earlier professor, who had been working on the atomic bomb project in the USA, and Marc Oliphant, whose team at Birmingham University had invented the high-power magnetron, the key component in the development of air-borne radar. Writing in the Harwell magazine sometime later, Cockcroft recalled:

> We thought we would have an establishment on a modest scale, with a pile (atomic test reactor) and a Van de Graaff machine (simple electrostatic accelerator) and a few other tools of nuclear physics. It should be near a university, there should be a small town for the staff and the countryside should be pleasant.

This slightly bucolic concept did not entirely survive the subsequent discussions and the Cabinet decided that the primary aim of the British effort should be directed towards all aspects of atomic energy, including the production of an "Independent Deterrent".[1]

Cockcroft was appointed Director of the new "Atomic Energy Research Establishment" (AERE) in January 1946, although he could not be released immediately from his work at Chalk River. The site chosen for the new laboratory was an RAF aerodrome at Harwell, in Berkshire, where the existing hangars and personnel accommodation could be used to give enterprise a flying start, and work started there in April.[2]

Although the new establishment's main job was to build reactors and carry out research on nuclear fission, it was also to have particle accelerators and it is this side of the work with which John was concerned. In 1950, the bomb work was transferred to the Atomic Weapons Research Establishment (AWRE) at a new laboratory built at Aldermaston in Berkshire, under the direction of William (later Lord) Penney. The production of nuclear fuels was centred at Risley in Lancashire and thenceforth Harwell was mainly concerned with basic research and peaceful applications.

Cockcroft, while still in Canada, decided that, in addition to the Van de Graaff he had earlier envisaged, a cyclotron was needed, primarily to produce radioactive isotopes and fast neutrons, but that provision should also be made to provide a beam of protons for high-energy particle physics research. He started to draw up a specification for such a machine, which uses strong magnetic fields to constrain charged particles, such as protons, into circular arcs, so that they can be accelerated repeatedly to much higher energies than obtainable with a single-pass accelerator like a Van de Graaff. In his absence, Cockcroft appointed Herbert Skinner from TRE to be his deputy at Harwell and head of the General Physics Division. To lead the Cyclotron Group in that Division, Skinner chose T.G. (Gerry) Pickavance, who had been running the Liverpool machine in Chadwick's absence, a brilliant young physicist who at one time was thought by his colleagues at Liverpool to be getting a bit above his station, which they demonstrated by fixing him spread-eagled to the floor with nails driven through his lab coat! Also, as we have seen, Skinner invited John to come and supervise the engineering aspects of the project.

The performance specification that was sent to Skinner by Cockcroft in February 1946 called for a machine to accelerate protons, deuterons (the nuclei of heavy hydrogen) and alpha particles (the nuclei of helium), to energies in excess of 50 MeV (Millions of electron-volts, the energy gained by an electron when accelerated by a potential difference of one million volts). It was proposed to adopt a design based on the machine which was being built at the Massachusetts Institute of Technology (MIT) by Stanley Livingston, who had assisted Ernest Lawrence, the inventor of the cyclotron, in the construction of his first machines. No sooner had the drawings for the MIT design arrived in Britain than news came of a new type of machine, the frequency-modulated cyclotron, or synchro-cyclotron, proposed almost simultaneously in the USA and the USSR, which overcame the relativistic effects that limited the maximum energy of particles in a cyclotron. It was then decided to build a synchro-cyclotron with a magnet pole diameter of 110 inches, the largest that had been considered in the previous studies. (Following American practice, the size of a cyclotron was determined by the size of the magnet pole in inches).

Skinner formed a Cyclotron Panel to discuss the design of the machine, which was then intended to be the first of several, to be located at Universities that were prominent in the nuclear physics field. Meetings were held at Cambridge and Birmingham, as well as at Liverpool University, where most of the theoretical work was done by Gerry Pickavance and the staff of the Physics Department in March 1946. At Malvern, John was joined by Mac Snowdon, who had been recruited by Skinner to work on the radio frequency system for the cyclotron. Mac recalled:

> John was a modest and serious minded young man, as I suppose I was at that time, and we got on very well right from the start, The small group of experimental staff brought together at Malvern worked very effectively and smoothly under John's direction. He was always available, sitting at his desk with pipe in hand, for discussion or advice. Later, when engineers and draughtsmen were engaged to work on the project at Harwell, the going was not so straightforward and he would have to spend many hours over their drawing boards putting across his ideas.

In April, the buildings became available at Harwell, so that the embryo team could move in. The facilities at Harwell were somewhat primitive at first, and there are many tales of the hardships of the first winter there. It was one of the coldest winters of recent years and there was a shortage of fuel, there were frequent power cuts and food was still rationed. John and Renie and their infant daughter were allocated a prefabricated bungalow which was very poorly insulated. When he couldn't get any coal for the stove, John used to go out at night and bring back one of the wooden posts which had been used to support the perimeter fence but which were then lying about the airfield. Being impregnated with bitumen, these burnt well after being cut up.

Once at Harwell, the group could go down to the detailed design and practical engineering of the cyclotron. That was where John came in. Although Pickavance was the head of the Cyclotron group, he was away in the USA for several months early in the project and it was left to John to organise all except the radiofrequency system, which was Snowdon's responsibility.

The construction of such a cyclotron was a large engineering project, by any standards, especially for a young man in his twenties. It involved building a 750 ton electromagnet, together with vacuum systems, power supplies, ion sources, water cooling systems and the necessary controls. The magnet was to be engineered and built by C.A. Parsons, a large electrical engineering firm, whose main product was turbo-generators for power stations. Most of the parts that could not be purchased from industrial sources were to be built in the Harwell workshops. John's earlier drawing office experience was put to good use in the intricate fitting together of these different machine components. The initial team assigned to build this monster (by the then current standards) consisted of three in the Scientific grade, Pickavance, Adams and Snowden, together with seven in the Experimental grade, at a time when John estimated that a total of 14 was the minimum. On accepting the post, John had been promised promotion to Senior Scientific Officer, corresponding to his new responsibilities, but formal appointment to this grade was delayed for six months while the administration argued about his qualifications.

Despite still being tied up in Canada, Cockcroft took a lively interest in the design, since he had built a small cyclotron at Cambridge before the war. This can be seen from John's notebooks, which are full of sketches and calculations for different magnets, including comparisons between the "Cockcroft magnet" and the Harwell one. Cockcroft also managed to get over to Harwell for a visit, bringing with him extensive data on the machines in the USA. The notebooks also show how John went into every detail of the design, doing his own calculations for things as diverse as particle trajectories in the machine and heat transfer to the cooling water.

Cockcroft came back from Canada in September 1946 and took over control of the establishment. He ruled over Harwell with a light touch,

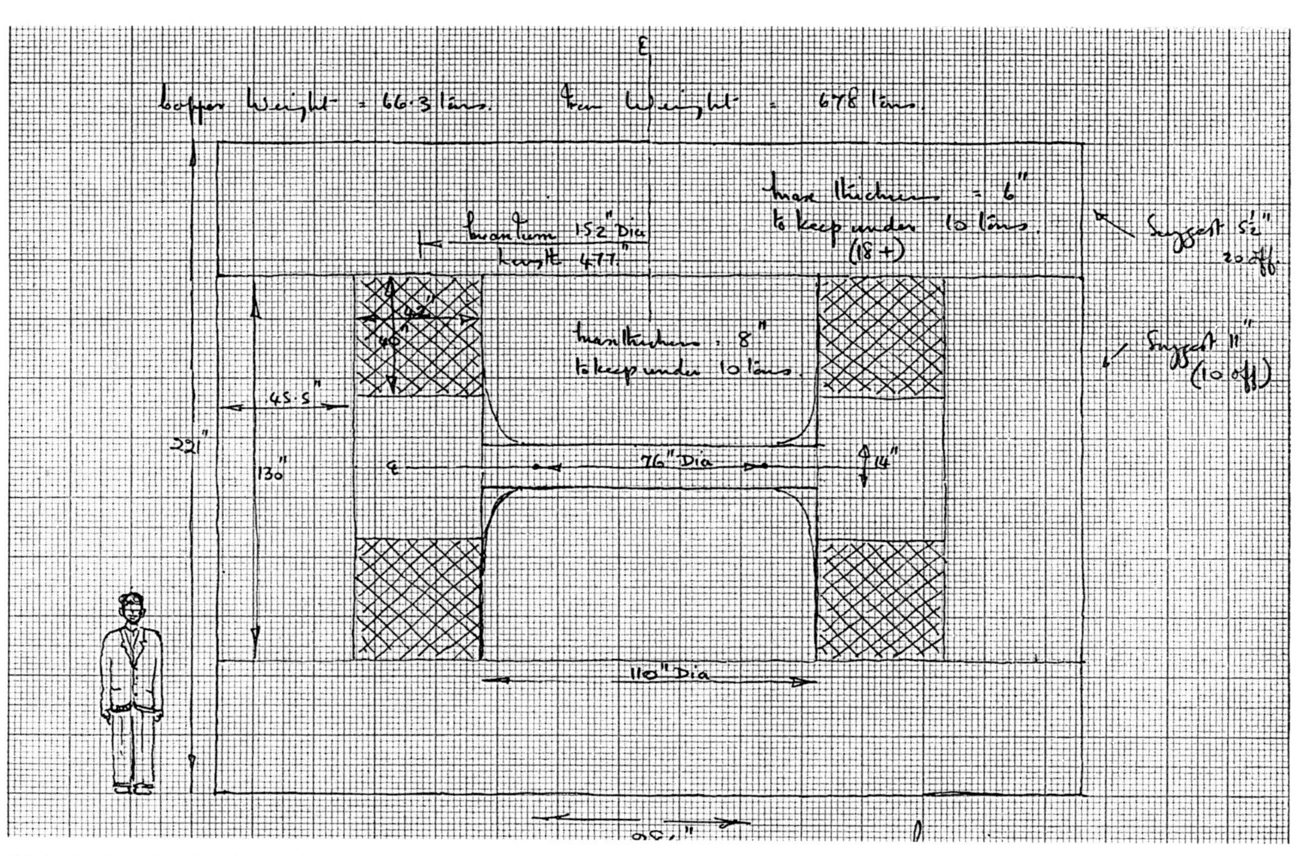

One of John's sketches for the cyclotron magnet.

believing that if his staff were engaged in activities that seemed worth while he should not interfere. He was convinced that future progress required the early recognition of talents in the young scientists and their every possible encouragement. John looked up to Cockcroft as to no other man, and always had large signed sketch of his chubby face on his office wall. On Cockcroft's death, John wrote:

> For me and for many of my generation in England, he was the great patron who provided the conditions in which we grew in stature as scientists and engineers and to him we owe our development. Without him we would, no doubt, have followed modest careers in government or industrial laboratories, but he pushed us to tackle jobs and projects which we never dreamt we could do or would ever have the opportunity to do and he sent us all over the world.[3]

Although the design of the synchro-cyclotron was almost settled, John did not like the relative proportions between the coils and the magnet yoke and worked out a design which would use the same coil cross-section but with the pole size increased to 130 inches. He estimated that this would add about £50,000 to the cost of the machine, but would allow the maximum energy to be increased appreciably. He made a wooden scale model to try to convince Skinner to accept the increase, but he was unsuccessful. The change was turned down on the grounds of cost, additional delay and that it would require the radio-frequency and other systems to be redesigned. It was a decision which was later found to have been unfortunate, as the maximum proton energy of the 110 inch machine was 175 MeV, just too low to produce mesons, transient particles with special properties, in useful quantities. This was not known at that time, though John had estimates for the variation of the rate of meson production with energy copied down in his later notebooks, as soon as the figures were available.

Skinner got a shock when the cost estimates for the magnet came in from Parsons, and the work was held up while these were investigated, but it was restarted when Skinner was convinced that they were reasonable. The estimates for the other components did not cause such trauma and the machine was completed in three years, the first protons being accelerated in December 1949. The final cost was within the original budgetary estimate of £130,000 made in June 1946. It was the second synchro-cyclotron in the world to operate, after the Berkeley 184 inch machine, which had a head start, with its pre-war magnet. The Harwell machine was then the highest energy accelerator in Europe, and it ran successfully for the next 30

The Harwell 110 inch synchro-cyclotron.

years, producing radio-isotopes for medical and other purposes in its later years. It was finally shut down for financial reasons, not for any technical short-comings. Many years later, in 1978, at the Commemoration Evening on the occasion of the shutting down of the Nimrod accelerator, it was remarked that "the little synchro-cyclotron behind the AERE fence that operated in 1949… has seen NIRNS and Nimrod come and go but still itself runs happily on."

John had not only been responsible for the engineering and coordination of this complex machine, but had also found time to advance the art in a number of areas, particularly in the methods of finding the central plane of a magnet and in beam monitoring and extraction devices. He had carried out this work at the rank of Senior Scientific Officer, his promotion to Principal Scientific Officer only coming through two years later.

The period following the completion of this project was not a happy one for John. He had been working at a high pitch and his project had come to a successful conclusion. Instead of having his day-to-day activities mapped out by the exigencies of the project, he had time to ponder on his future career. He was assigned to a reactor physics group and was told or got the impression that he would be asked to work on a fast reactor project. He thought that his lack of nuclear physics training would be a handicap he could not surmount.

His wife, Renie, wrote of this period:

> In his usual way he wanted to understand this new field and started to work night and day to study all the reports. Whether he had worked too hard or whether he could not work without being allowed to exercise his own initiative I do not know, but he became very unhappy and depressed and suffered a nervous breakdown. He did not want to talk to anyone except to me and thought himself a complete failure. I think too at this time the long period of food rationing may have taken its toll.

Her solution was to send him off for a week to stay with his uncle, who was a pork butcher. After being fed like a king, he returned somewhat better, but could not sleep. Renie's reaction to this was to order a load of logs for him to saw until he was physically tired! This therapy seemed to work, and after a few weeks he asked to see Cockcroft to discuss his future career.

Cockcroft was very sympathetic and it was decided that John would work on the development of a high-powered klystron, a microwave power generator much more powerful than the magnetron which had been the mainstay of the wartime radar systems. For this new development, he was to collaborate with Mervyn Hine, whose hesitant approach and mild appearance hid a keen brain and a sometimes malicious tongue. He had just gained his PhD in nuclear physics at Cambridge and decided to take up a vacancy at AERE rather than "become a lecturer at a second class university". Hine, never one to mince words, said:

> It was suggested that I should work with John on this new project of copying the Stamford 20 MW klystron, because Cockcroft had been asked by the Ministry of Defence "for God's sake do something British", since the radar valve industry was incapable of doing anything on the level of the Americans and something must be done.

This project was the result of the outbreak of the Korean war when pressure was put on Harwell to release some people to go back to the Ministry of Defence. This was resisted and an agreement was reached to do some defence work at Harwell instead. After eighteen month's work, a successful klystron was produced. It seemed that by then John had regained his equilibrium and his enjoyment of life. This period was the beginning of a life-long friendship with Mervyn, despite some disagreements as to what was the best for CERN in later years.

During this period at Harwell, Renie had given birth to a second daughter, Katherine, on the 4th December 1947 and a son, Christopher, on the 15th May 1951, so completing the family. By this time, Harwell had built some houses in the grounds of an old manor house on the outskirts of Abingdon, and John was allocated one of these, which was luxurious compared with the "pre-fab" they had previously been living in.

Before following John's career further, we must take a look at what had been happening on the European scene, as that was to have a major impact on his future.

References

1 Gowing, M. *Britain & Atomic Energy — 1939 – 1945.* London. MacMillan, 1964.
2 *Harwell, 1946 – 1951.* H.M.S.O. 1952.
3 Hartcup, G. and Allibone, T.E. *Cockcroft and the Atom,* A. Hilger, 1984.

Chapter 4: THE EUROPEAN ELEMENT

In a talk given in 1982, John pointed out that European physics had enjoyed a fruitful period before the war when the physics of the atom was unravelled and a start made in exploring its nucleus. After the war, there seemed to be no way of getting back to the forefront of this research again or even re-starting the research in many of the European countries. Most of the accelerators were small and none could match the giant machines being built in the USA. (The higher the energy of the accelerator, the more powerful the "microscope" it provides for investigating the fundamental properties of matter).
He went on to say:

> By 1950, it was already clear that no single country in Europe was willing to pay the price for getting back again to the front line of this research. It might have declined irretrievably had it not been for a quite different group of people, not scientists at all but statesmen, who were looking for some way of demonstrating that the European countries could achieve far more together than they could ever do separately. It was the coming together of these two groups of people, the nuclear scientists anxious about the future of their research and the statesmen interested in joint European action, that saved nuclear particle physics in Europe. The reason the two groups came together was again due to the war-time nuclear developments.
>
> Just after the war, the United Nations Organization set up a commission on the control of nuclear energy, the members of which were government officials and diplomats. They in turn called in a number of eminent nuclear scientists as experts, many of whom were the nuclear physicists who had worked on the war-time developments and were by then in leading positions in the Atomic Energy Agencies. After the official meetings were over for the day, these physicists talked with the officials about the future of scientific research in Europe and from these conversations emerged the idea of inter-governmental action. Nearly all the pre-war generation of nuclear physicists got involved in these discussions at one time or another, so it is hardly surprising that nuclear particle physics was chosen as the first joint European action. It also had the merit that no commercial application was foreseen for the research and no military purpose – two features which made it more acceptable to governments.

The story of the formation of the organisation that eventually sprung from these discussions has been recorded in detail [1]. In December 1949, a European Cultural Conference was held in Lausanne where it was proposed that the possibilities for the creation of a European institute for nuclear science should be explored. Six months later, at the annual conference of UNESCO held in Florence, the American Nobel Laureate, Isidore Rabi, proposed the formation of one or more regional European laboratories, including one for nuclear science. Coming from such a prominent American physicist, co-founder of the Brookhaven National Laboratory near New York, this proposal carried considerable weight, as well as showing that the United States, or at least some of its top scientists, were no longer against collaboration on nuclear matters, and it was adopted by the General Assembly of UNESCO.

In the months following this conference, many discussions took place around Rabi's proposal. Pierre Auger, head of the Exact Sciences Department, set up a special office in UNESCO and got together a number of physicists to help him with the planning of a new European laboratory. His closest collaborator was Eduardo Amaldi, then the Professor of Experimental Physics at the University of Rome, a little human dynamo who was also an ardent European. By December 1950, the idea had crystallized into a project for a laboratory centred on a giant accelerator. Not everyone was in favour of this. Niels Bohr, the Danish theoretician, famous for his work on the structure of the atom, was not very enthusiastic about the construction of enormous equipment for nuclear research and later proposed that the new laboratory should be located in Copenhagen and carry out experiments on existing machines, such as that at Liverpool.

Up to 1950, no British scientists or administrators had been officially involved in these activities, although the British community of scientists knew what was going on. The main reason for this lay in the fact that, at that time, Britain was the leading nation in Europe in the field of nuclear physics. This was largely as a result of the experience gained by her scientists during the war, together with the political will and financial means to launch a national atomic energy programme which included the construction of a number of accelerators, as well as the reactor and bomb development. To supplement the 175 MeV synchrocyclotron at Harwell, which had been the highest energy machine in Europe for a while, even more powerful machines were under construction at three universities: a 300 MeV electron synchotron at Glasgow, a 400 MeV synchro-cyclotron at Liverpool and a 1000 MeV proton synchotron at Birmingham. The synchrotron was a development of the synchro-cyclotron which used an annulus of smaller magnets instead of the one massive magnet, thus reducing the cost of a machine of given energy. The time and money already committed to this national programme raised doubts in many minds as to whether Britain should participate in yet another accelerator programme. In particular, the Nuclear Physics Committee of the Department of Scientific and Industrial Research (D. S. I. R.), chaired by Chadwick, noted early in 1951, that "there is a very definite balance of opinion that this country should not join directly in the establishment of such a Laboratory and should not promise support with either men or money". However, the official response indicated that if the European Laboratory was established, Britain would be prepared to help "by advice and assistance on equipment and its design, possibly by seconding men for short periods and by accepting men from the laboratory for training experience and research on our own projects".

The role of Cockcroft, as Director of AERE, where he had built up a team of expert accelerator builders, was important. He agreed, as most British physicists did, that the Laboratory was a good thing for the future of physics on the continent and he took positive steps to help his European colleagues to formulate a realistic programme. When, in May 1951, Pierre Auger convened a meeting of experts in Paris to prepare a proposal to be put to a UNESCO meeting later in the year, Cockcroft sent Frank Goward, who had built the first electron synchrotron at TRE Malvern in 1946 and had been responsible for other developments in this field, before he was transferred to Harwell in 1951. The attitude of British physicists changed steadily during 1951 and in this the long-standing personal ties between Cockcroft, Amaldi and Auger played an important part.

On the 15th February 1952, as the result of further meetings, delegates from eleven countries signed an agreement in Geneva "constituting a Council of Representatives of European States for planning an international laboratory and organizing other forms of cooperation in nuclear research". When the "instrument of ratification" was deposited in at UNESCO House in Paris on the 2 May 1952, the provisional *Conseil Européen pour la Reserche Nucleaire* (CERN) came into existence, and Amaldi was appointed Secretary-General.

It had been decided that the new laboratory would build two accelerators. The first would be a 600 MeV synchro-cyclotron, which could be built rapidly, using well-known principles, to give an early start to the experimental programme. This would be followed by a proton synchrotron of at least 10 GeV, to give it a significant advantage over any machine known to being built or planned. (A 10 GeV machine was to be built at Dubna, in the USSR, but that was not known at the time). The Council of this new organization first met in May 1952, in Paris, and it decided to set up four study groups; the Synchro-cyclotron Group under C.J. Bakker, a prominent Dutch physicist, the Proton Synchrotron Group under Odd Dahl, the Laboratory Group under Lew Kowarski, who had been one of the pioneers of the nuclear fission work that led up to the atomic bomb, and the Theoretical Group under Niels Bohr, who had somewhat softened his previous attitude against large machines.

At the third meeting of the Council, it was decided to locate the new laboratory near Geneva and a suitable site was sought. Although Britain was only officially in the position of 'observer', Sir Ben Lockspeiser, then head of the Department of Industrial and Scientific Research (DISR), was appointed to be Chairman of the Finance Committee of the provisional CERN.

John with Sir Ben Lockspeiser.

Odd Dahl was Professor of Physics, at the Chr. Michelson Institute in Bergen, Norway. Renie described him as a charming character and a good storyteller. He had been on one of Amundsen's expeditions to the North Pole as a pilot and had to assemble the aircraft out of a number of crates of parts and found that there were no instructions, so he had to work it out for himself! His Group was given the brief "to study particle accelerators for energies greater than 1 GeV (1000 MeV), and in particular the problems of building a scaled up version of the Cosmotron" (a 3 GeV synchrotron being built at Brookhaven in the U.S.A.). Dahl included Frank Goward from Harwell in his Group, which was made up of prominent physicists from Member states. It also included Dahl's assistant Kjell Johnsen, a tall, bluff, abstemious Norwegian with a passion for keeping fit. The group's work was centred at Dahl's Institute, but some of the members spent most of their time in their home institutions, where they could involve others in the work.

Since it was originally intended to build a scaled-up version of the Cosmotron, Dahl and Goward went to Brookhaven in August 1952 to make a detailed study of this machine, which was nearing completion. They were accompanied by another member of the Group, R. Wideröe, famous for his work on the betatron, a type of electron accelerator. During their visit, the discovery of the principle of alternate gradient 'strong focussing' was revealed.[2] This seemed to give the possibility of a drastic reduction in the sizes of the magnets and vacuum chambers needed for a given machine, compared with those of conventional design, which used 'weak focussing'.

Dahl seized on this new development with enthusiasm, seeing it as a way to get a more powerful machine for the same money. The team was set to work on the outline design of a synchrotron using the new strong focussing principle (colloquially referred to as "AG") and, after some frantic effort, a proposal was put by the PS Group to a meeting of the Provisional Council, held in Amsterdam in October 1952. This was to stop work on the 10 GeV weak focussing machine and instead concentrate on a strong focussing proton synchrotron of the highest possible energy, between 20 and 30 GeV, compatible with the same total cost.

Commenting on this decision, at a meeting called to pay tribute to Dahl's 70th birthday, John said:

> Undoubtedly this attitude of Dahl, so very characteristic of the man, largely determined the future success of CERN. It would have been so much easier to have advocated the safe course and to persuade the

> Council of CERN to accept it, but had CERN gone on to build a 10–15 GeV scaled-up Cosmotron, it would probably not have attained its present high standing in the world of elementary particle physics. It was also a very unselfish attitude for Dahl to take, because the whole nature of the Proton Synchrotron Group's work changed thereafter. Instead of being essentially an engineering group scaling up an existing machine based on well-established principles, it became a physics group studying the theory of accelerators, and only later did it return again to engineering design. Dahl... saw his function largely as a mechanical engineer leading an engineering group, but as a result of advocating the new focussing principle he thereby made himself responsible for a quite different kind of project.[3]

In contrast with this innovative approach, when the National Institute for Research in Nuclear Science (NIRNS) was formed in Britain in 1957, it was decided to play safe and build a 7 GeV weak focussing synchrotron, called Nimrod, on its site near Harwell. John, although by then committed to the construction of the CERN machine using the new ideas, supported this decision, as he still had at the back of his mind the possibility that some snag might crop up to prevent the CERN PS from working and he thought that a conventional machine could be a useful back-up for European physics in such a case. The thinking at the time was summed up in a poem attributed to a prominent physicist, Rudolf Peierls, which included the lines "For Awful Gamble stands AG — But if it works or not we'll see".

However, Nimrod was not finished until 1963, four years after the successful first operation of the PS. It was not only on this side of the Atlantic that other people were cautious, and the weak focussing Zero Gradient Synchrotron (ZGS) was started at the Argonne Laboratory in 1959 and did not come into operation until September 1963.

Towards the end of 1952, there were debates amongst the British physicists and scientific administrators about whether it would be better to have their own new machine or to join CERN. It was by now certain that CERN would go ahead even if Britain made no contribution. The rapid acceptance by the Dahl group of the strong focussing principle meant that if Britain did build its own big machine it would be smaller and come later than the CERN one. Cockcroft called for a meeting of Chadwick's Advisory Committee to try to come to a decision. This was held at Cambridge in November 1952 and Chadwick must have reversed his earlier objections, since the minutes record a unanimous call for Britain to join CERN. Now that the physicists spoke with one voice, they were taken more seriously by the Government, but it still remained to convince Lord Cherwell, Paymaster-General and scientific adviser to the Prime Minister, Winston Churchill, who was anti-European and would have preferred Britain to have its own machine. As we shall see in the next chapter, Amaldi came over to try to convince him, but he wrote later:

> Lord Cherwell appeared to be very clearly against the participation of the U.K. in the new organization. As soon as I was introduced in his office he said that the European Laboratory was to be one more of the many international bodies consuming money and producing papers of no practical use. I was annoyed and answered rather sharply that it was a great pity that the U.K. was not ready to join such a venture which, without any doubt, was destined to full success, and I went on explaining the reasons for my convictions. Lord Cherwell concluded the meeting by saying the problem had to be reconsidered by His Majesty's Government.

The interview lasted only ten minutes! Despite the impression he gave to Amaldi, Cherwell had already been persuaded not to oppose Britain joining, as long as some changes were made to the proposed Convention for CERN. Sir Ben Lockspeiser took an active part in the drawing up of this Convention, to establish a 'European Organization for Nuclear Research', which was presented to the Council at a meeting on the 1 June 1953, in Paris. He had done his work so well that the United Kingdom was the first of the nine nations to sign the Convention. Ironically, France and Italy, who had been forerunners in the establishment of the Provisional CERN, were the last to sign, due to administrative difficulties.

It may or may not have been in the minds of those who chose to adopt the new names (*Organisation Européenne pour la Recherche Nucléaire* in French) but to keep the acronym CERN, which did not now correspond to the name in either language, but this decision avoided having to have two acronyms, a situation from which most of the International organizations in Geneva suffer. On the other hand, the choice of the word "nuclear" in the title was unfortunate, as trouble was caused in later years by people who only think of that word in connection with nuclear power and nuclear weapons.

Some time later, in 1965, John wrote:

> During the period between the signing of the Convention on 1 July 1953 and its ratification on 24 September 1954, the organization had no legal existence and all work should have stopped, since in the event of insufficient signatories it would all have been a waste of time and money. In fact, the Interim Council and the Study Groups continued to plan the laboratory and to design its apparatus. The credit for this unorthodox action, which resulted in an enormous saving of time, must go to the many highly placed diplomats and administrators who risked their reputations on the eventual success of the venture, and to the young scientists and engineers who left secure positions in established laboratories in their own countries to lead a pioneer existence with their families in a foreign country, with no more than a 'moral commitment' to fall back on if it all failed[4].

However, the parliaments of all member states did ratify the Convention and CERN came officially into existence. Sir Ben Lockspeiser was elected the first Chairman of the CERN Council and Amaldi was offered the post of Director-General but refused, as he wanted to get back to his physics work at Rome. After some hesitation, the Council elected Felix Bloch, a Swiss theoretical physicist who was then working in the United States, to be the first Director-General. Bloch insisted that he should be able to carry on his original research work on magnetism at CERN and imported his equipment and two assistants at CERN's expense, as he seemed to think that the post of Director-General was just a nominal one. He did not find the administrative side to his liking and deputised it all to Amaldi, who found he had to spend most of his time in Geneva instead of getting on with his physics work at Rome.

After a year, Bloch resigned and the Council appointed Bakker, who up to then had been running the synchro-cyclotron (SC) group, to take his place. Amaldi returned to Rome, but was destined to continue to play a major role in the future of CERN, starting with his appointment as Chairman of the Scientific Policy Committee in 1958. As we will see later, Bakker himself was not all that keen on taking administrative decisions. From the beginning of CERN, the physics community has always insisted that the Director-General should be an eminent physicist, but eminent physicists are not necessarily good managers and administrators and it might have been better to have had a bipartite leadership of Chief Scientist and Managing Director.

Meanwhile, Dahl's Group was working on the new design for a 30

GeV proton synchrotron, three times the energy of the original project. Since most of those concerned could only spend part of their time on the project at this stage, Dahl formed semi-independent subgroups to look at different aspects. Although Britain was not yet officially taking part in the project, Cockcroft gave Goward the facilities to form a subgroup at Harwell, to study the particle orbits in the machine.

With this introduction to the European scene, we will now see how John became involved in this project.

References

1 Hermann, H. *et al., History of CERN, Vol. 1: Launching the European Organization for Nuclear Research.,* North Holland, 1988.

2 Courant, E.D., Livingston, S.M. and Snyder, H.S. "The strong-focussing synchrotron – a new high energy accelerator." *Phys. Rev.* **88** (1952) 1190-1196.

3 A.S. John Griegs Boktrykkeri, *Festskrift til Odd Dahl,* Bergen 1968.

4 Cockcroft, Sir John (Ed), *The Organization of Research Establishments,* Cambridge University Press, 1965.

Chapter 5: JOHN AT CERN

As we have seen, at the time when officialdom in Britain was reluctant about joining CERN, Cockcroft invited Edoardo Amaldi to come to Britain, in December 1952, to try to convince Lord Cherwell to abandon, or at least to mitigate, his negative attitude towards this new continental initiative. In his agreement to this visit, Amaldi told Cockcroft that he would like to meet some young physicists or engineers who might be interested in participating in the construction of the new European Laboratory. In the published version of the John Adams Memorial Lecture given at CERN in December 1985, Amaldi wrote:

> When in London Sir John (Cockcroft) had asked me to join him for lunch, he mentioned that his arrangements also aimed at giving me the opportunity to meet a young engineer who had worked on the construction of the Harwell synchro-cyclotron. In a few words, typical of his style, Cockcroft succeeded in communicating to me his confidence in the capabilities of this young man and that no objection would be raised from the British side by an offer from CERN, since the Harwell programme did not foresee, in the near future, developments requiring his specific abilities. At lunch, I met John Adams. I was immediately impressed by his competence in accelerators, his open mind on a variety of scientific and technical subjects and his interest in the problems of creating a new European Laboratory. [1]

After his session with Cherwell, Amaldi was driven to Harwell by John Adams and "the conversation during the three hours drive to Harwell confirmed my first impression; he was remarkable by any standard and he was ready, incredibly ready, to come to work for CERN". At dinner, and afterwards, Amaldi had interesting conversations with members of the General Physics Division, including Donald Fry, who had taken over the Division when Skinner returned to Liverpool in 1950, Frank Goward, John Lawson, who had worked with Goward at TRE on the electron synchrotron and then joined Harwell to build a new kind of klystron, and Bill Walkinshaw, interested in the theory of accelerators. Amaldi continued:

> I was also very impressed by the other young people I met at Harwell, not only because of their scientific abilities but also because of the detailed information they had about CERN. Contrary to the impression

> that I had got from Lord Cherwell early in the afternoon they were not at all insularly minded. Clearly Goward was, in great part, responsible for the rather complete information that all of them had about CERN.

This visit took place about two months after the decision had been taken to change the design to use the new strong focussing principle and Amaldi noted that "all the young people I met at Harwell were very interested in this principle and its implementation, which was clearly at the centre of their thoughts".

During this period, according to Mervyn Hine, he, John Adams and John Lawson, although not officially members of Goward's subgroup that had been assigned to work on the European accelerator, often joined in the meetings and discussions, to the extent that they became accepted as collaborators and unofficial members of the team. The trio discussed the strong focussing principle at length and began to see reasons why it would not work in the form proposed, with very high 'n' value (a measure of the focussing strength) and very small apertures for the beam. Lawson pointed out that small misalignments and inhomogeneities in the guiding magnetic field would result in a build up of the oscillations of the beam, resulting in its total loss before the full energy was attained. Luckily, this was over pessimistic; as Lawson admitted "I was pressing this point hard because I thought everyone was going forward with too much gay abandon on this subject. I was rather over cautious about that; I always tend to be." However, it meant that the 'n' value had to be reduced significantly, from several thousand to a few hundred, and very tight tolerances placed on the magnet properties and position. The result of this was that, while the size of the magnets and vacuum chamber could be reduced significantly compared with the previous weak-focussing machines, the reduction could not be as great as had been originally thought. The 'n' value was reduced by stages from the original 4,000 to about 400, which meant that the total magnet weight went up from the original 800 tons to 4,000 tons, for the 30 GeV machine, but this was still small compared with the 36,000 tons of the magnet for the Dubna 10 GeV weak-focussing machine. The injection energy was also increased from 15 to 50 MeV, since the magnet inhomogeneities are greatest at low fields.

Hine pointed out the problem of "stop-bands", when the number of oscillations of a particle going round the machine was equal to the ratio of two integers. This also meant a further tightening of the tolerances on the magnetic fields and the positions of the magnets. Since the orbit theory involved complex non-linear equations, the help of the National Physical Laboratory was sought, as this laboratory had what was then the largest computer in Britain which was available for scientific work, and additional resonances were detected. With all these difficulties being discovered, it was said that "their Jeremiah-like prognostications about inhomogeneities became a depressing feature of the Group meetings." At this period, the abilities of John and Mervyn Hine were so complementary, they worked so well together and were almost exactly the same age, that they became known as the Harwell twins and sometimes the "terrible twins" in view of their disruptive activities. Lawson reckoned that Mervyn had a better background in mathematics but John was very good at picking up the essential points and actually doing something about it. In some respects, John was the dominant one. A meeting was held at Harwell at the end of 1952 at which, according to Mervyn Hine, the first set of realistic parameters was discussed. He told me "I remember, at the end of it, John summarised and took over the meeting at that stage. It was very interesting; I was prepared to do it and then I saw that John had stepped into the authority position and he wrote a summary on the blackboard in his wonderfully clear left-handed writing". It was at this meeting that Kjell Johnsen first met John, an association that was later to lead to one of the few long-standing conflicts of personality in John' s life.

As a result of this work, Adams, Hine and Lawson published a paper which showed that the resonances could be managed if sufficiently tight tolerances were kept on the magnetic field and gradient in each magnet, and in the position of the magnets. [2]

Although a site for the new laboratory had been chosen just outside Geneva, it would be some time before any, even temporary, buildings could be put up, to allow the Group to be brought together. A meeting was held in Geneva in September and Hine remembers that he and John were doing the calculations for the latest design in an hotel bedroom. The situation was eased a little when, through the kindness of the Geneva University, space was found in the Institute of Physics and the nucleus of the Group moved there in October 1953. This included John and Hine, who became officially full-time staff members of CERN the following month, on secondment from the AERE. Dahl himself did not move to Geneva, but directed the work from Bergen, appointing Goward to act as his deputy in Geneva .

Renie and the children followed John about a month later when a flat was found for them in Geneva. They were lucky to find places for the two girls at the International school, where there was normally a waiting list, but two became available when Mrs. Guy Burgess decided to join her husband in the USSR and take her two sons with her. John and Mervyn had agreed to go to CERN at effectively the same salary as they had received at Harwell, but they found the cost of living, especially that of renting an apartment and the school fees, considerably higher. Luckily for them, the Cockcrofts came over to Geneva for a holiday the following spring and Renie Adams went with Lady Cockcroft to do the family shopping. After having to borrow money from Renie to pay for her purchases, which cost vastly more than she had expected, Lady Cockcroft brought home to her husband that the cost of living in Geneva was at least double that in Britain, so that Cockcroft proposed, at the next Council meeting, that the CERN salaries should be raised and that the increase should be back-dated. This was agreed and the resulting windfall enabled John to buy his first car in Geneva, a second-hand Standard Vanguard.

As soon as the Group was established in Geneva, a conference on the Alternating-Gradient (Strong Focussing) Proton Synchrotron was held there from the 26th to 28th of October. Presentations of the work carried out by the subgroups were given and a set of consistent parameters for a 30 GeV machine were presented by the PS Group. Of the Harwell subgroup, Goward gave a paper on the design principles, Hine one on nonlinear orbit theory and Adams on a design based on the linear theory. The orbit papers were completed by one on another problem of strong-focussing machines, that of phase oscillations and the possibilities of beam loss on transition, which was given by Kjell Johnsen.

Previously, Dahl had invited two American experts, John and Hildred Blewett, to come over from Brookhaven, where they had made major contributions to the Cosmotron and to the design for the 30 GeV Alternating Gradient Synchrotron (AGS), which was to use the new principle. However, Brookhaven decided to carry out some model work to prove that this principle would work before committing themselves to its adoption for their new machine, and so the Blewetts came to Europe to help with the design of the CERN Proton Synchrotron (PS). Hildred Blewett edited the published version of the talks given at this conference [3], in which she wrote:

> This design is not to be construed as final, but rather as the first in a series of designs that gradually converge as the many conflicting fac-

tors are balanced against each other and as more model components are constructed and measured.

This so exactly mirrored John's attitude to project management that it seems likely th[illegible] Mildred's words were a reflection of some statement he had mad[illegible]

The [illegible] resulting from the October conference were discussed at the fol[illegible]RN Council meeting and dismay was expressed at the rise in th[illegible]cost. Dahl was instructed to look at the effect of reducing the e[illegible]GeV. This was done and a new proposal put to a session of th[illegible]ld in March 1954 and it was accepted. The magnet weight w[illegible] 3,300 tons and the 'n' value to about 300.

It se[illegible]r Dahl did not take much interest in the administrative side, [illegible]ok it on himself to make sure that he was *au fait* with every[illegible]t on, since one of his notebooks has a list, dated 15 Novembe[illegible]fter he had arrived in Geneva, which gives the names, ag[illegible]salaries of the then present staff, which numbered 19, includi[illegible] Blewetts. There was also a list of about 18 applicants and a [illegible]nary of the situation at 30 November and forecasts for th[illegible]nths. According to Hine, neither he nor John were attache[illegible]ular part of the machine but were both on the general desig[illegible]

The situ[illegible]ged abruptly when Frank Goward was found unconscious [illegible] the physics building. He was immediately flown back to [illegible]ain tumour was diagnosed and he died in March 1954. La[illegible]gh opinion of Goward's intellectual ability and scientific o[illegible]his was not shared by John and Mervyn. Lawson says that [illegible]e antipathy between Goward on one hand and John and Me[illegible] other and he thinks that in the later years Goward was already suffering from the illness that killed him and perhaps did some things that were a bit tactless and did not always show good judgement .

Dahl appointed John to act as his deputy in Geneva but was having second thoughts about moving there himself, as his wife was rather frail, and he submitted his resignation in October. This was a double blow to the project and it was necessary to appoint a successor to take over the PS Group. The Council considered Christoph Schmelzer, who was the most senior in age in the Group, but decided to appoint John Adams to this position.

On this decision, Kjell Johnsen told me:

> For us people in the group, we considered that John was the most able person within the group for doing that kind of thing; his rational approach, his systematic approach, his calm attitude and so on made it natural, as well as his knowledge from the cyclotron work and from the radar work. This made him rather impressive to many of us and he seemed the natural choice to us. But it was not considered unnatural that Schmeltzer was a candidate in peoples' minds. He never radiated the same confidence in what he did and he did not show as much over-all knowledge as John did, but he was quite a few years older. It was Odd Dahl who pressed the view of the group to Council the most strongly, as he had also made up his mind fairly early that this job was for John. He had the experience of big projects and most of the others were straight from university. What we must not forget is that many of the young people we met in the accelerator field from Britain and America had done a tremendous job during the war and we had the highest respect for them. It showed that young people could take on these responsibilities if they were really asked to do so. These people came out of the war, young, but with remarkable experience. They showed confidence but also other people had confidence in them. That is probably why, even to us, the fact that John was less than 35 years of age at this time didn' t bother us.

On taking over the project, John had not only to leave the realm of the abstract design of the machine to deal with the hard reality of getting the machine built, but also he had the problem of welding together a staff of about 50 disparate people from many nations into an effective team. He started his Monday morning "Parameter" meetings, where everyone responsible for a particular part of the machine was expected to give a progress report, followed by discussions on current problems. Hine recalled:

> There were some difficulties; the sort of professional attitude of "that's my part of the job and I go away and do it and I don' t have to talk to anyone about it. " The idea of everyone telling the others what they were doing once a week was a little strange, but they did fall in fairly rapidly and the Parameter meeting became fairly automatic and people slotted into their positions. John had the policy at first of trying to put two people on to any important job, so there was some back-up. From then on it was a fairly smooth running organization.

Amaldi joined in some of these meetings and he wrote:

> I was very much impressed - as everybody else - of John Adams style and efficiency and also of the role played systematically by Mervyn Hine whose function was that of pulling together the values agreed for the various parameters ... to be sure that all of them were fitting in a coherent scheme. All the people that have participated in those meetings still recall with great admiration John Adams as director and leader of these vital discussions.[4]

Despite these remarks, it seems that there was still some tension within the Group. According to Johnsen this was not enough to interfere with the work, but was due to the attitude of the British. He thought it was the result of the war work and the fact that Britain had more accelerators than any other European country and greater experience in building them. He said

> There is no doubt that John and certain other people had a tendency to believe that the British knew better than the other ones what to do about this and that. A British colleague of mine once said to me: "The British don't just believe they are the best; they know it". It was true of this period. I felt that John had difficulties in trusting some of the other people.

This attitude must have been apparent outside the Group. It seems that Lew Kowarski, who was then head of the Scientific and Technical Services Division, when asked what was going on, replied that the PS Division was two different things; "one is a sort of middle European professor, Hine, and the other is a well run British firm. " Hine says that everyone thought that he was John' s private assistant and completely in his confidence, but that was not true. "Technical matters he would certainly discuss with me, but he kept to himself many managerial problems". John was certainly appreciative of his help. Robert Lévy-Mandel, visiting from the Saclay Laboratory near Paris, said about John at this time, "He admitted freely that he did not know everything about the subject, but he said 'if I don't know I knock on the wall' and next door was Mervyn's office. "

Those who participated say that, despite these difficulties, what impressed them most was the terrific enthusiasm the project generated. All felt that they were taking part in something really worth while, opening new frontiers to physics and engineering. In John' s case, the enthusiasm was kept muted. While he spread an atmosphere of calm assurance inside the meetings, outside them John would give the impression to many that he was deeply pessimistic, since he always had at the back of his mind that something that had been overlooked would crop up to prevent the machine working as foreseen. He would always try to identify possible difficulties and search for ways of overcoming them, minimising the risk by building up a second line of defence wherever possible. He would much rather seem pessimistic and be proved wrong than show undue optimism.

It is not the purpose of this biography to give technical details of the PS, but some new techniques were introduced in its construction. Because of the tight tolerances on the magnet position, it was decided to mount them on a reinforced concrete ring, which was supported on piles which went down to the underlying rock, and to control the temperature of the tunnel within close limits. The reports issued in this period show into what detail John went into the geology of the area to avoid unpleasant surprises. Another precaution was taken. To avoid differences between the magnets due to variations in the properties of the separate batches of the sheet steel magnet laminations, no magnet cores were assembled until all the laminations were manufactured and then they were mixed thoroughly, so that each magnet had laminations from most of the batches. After that, the laminations were to be glued together, using an epoxy resin, to make the magnet cores. The contract for this work was awarded to a consortium of two Italian firms. John gave Giorgio Brianti, a bright young redheaded Italian who had recently joined his group, the job of resident inspector, to ensure that the work was done correctly. Brianti remembered this period:

> At the beginning of the production of the cores, most of them were out of tolerance. This was not because of any particular difficulty. but because of carelessness and the desire of the firms to spend the minimum. So I went into dispute with the management and said "If you continue, you do it at your own risk, because we will never accept them." It was a hard decision for someone as young as I to take, but it was fully backed up by John, without further investigation. You had to demonstrate to him that your work was alright, and then he trusted you. I was very impressed by his behaviour. It is part of the quality of being a leader.

John's idea of being a leader extended beyond the construction of the PS, to the preparation for the experiments, or rather the lack of it, and to the administration of the laboratory, as we shall see in the next chapter, but here we will continue to follow the progress of the accelerator.

The building of the machine proceeded at a great rate, and many firms in Europe were stretched to the utmost to meet the stringent requirements for most of the components. For example, the firm I was then working for, Metropolitan-Vickers in Manchester, where Cockcroft had served his apprenticeship, had secured the contract for the manufacture of the major components of the 50 MeV linac injector, mainly because the firm was also building a similar machine for Harwell. The basic design work was done as a collaboration between the Metropolitan-Vickers Research Department Accelerator Section, of which I was a member, and delegates from Harwell and CERN, but the detailed engineering and manufacture were carried out in a factory that was more at home building big steam turbo-generators than linacs, with the results that many of the early components that were made did not meet the specifications and had to be rejected and new ones made. It took some time before a full set of satisfactory parts could be gathered together to be sent to Geneva to be assembled on the site.

With a lot of work from the CERN staff, satisfactory components began to roll in, and the accelerator tunnel and associated buildings were completed, so that the work of installation could start. As the completion

The magnets in the PS ring tunnel. The proton beam from the injector comes in from the right.

Mervyn Hine and Kjell Johnsen at the PS controls.

date grew near, even though John was fully occupied by the vast number of day to day things that the leader of a big project has to deal with, it was never far from the back of his mind that something might go wrong. After all, this was the first proton machine using the new principle of strong focussing, where the protons had to go through a period during the acceleration cycle when phase stability was lost, the transition period. Would this result in a massive loss of protons? Also there were the resonances that he and Hine had worked out; would it be possible to keep between them? There were theoretical studies that showed that these problems should be solvable, but had any factor been left out of the equations? The decision to change the design to a strong-focussing machine had not been made by John, but he had taken the job on to build such a machine and it was his responsibility to deliver the goods.

When all the major components were in place and tests started, there was some initial optimism, but that soon turned to frustration. Early in September, a beam of protons was injected into the PS and had gone one turn round the circle of magnets, an event that John was able to report to the International Accelerator Conference that was taking place at CERN at that time. When the radio-frequency equipment was ready, early in October, attempts at acceleration were started, but without much success. The beam could be captured but it was lost after a few milliseconds. The focussing seemed wrong and there were a number of unexplained effects. The weeks went by, with the last work in finishing the machine being carried out during the day, and tests during the night. A number of things were found to be wrong and were put right, but still the beam was lost after a few milliseconds. The day when a beam was finally accelerated further was described vividly in an article written for the CERN Courier [4] by Hildred Blewett, who had earlier gone back to Brookhaven but had now returned for the Conference and stayed on to see the start-up of the PS. She wrote:

Wondering why it didn't work! Left to right - John, Geibel, Blewett, Lloyd Smith, Schmelzer, Schnell, Germaine.

> Remember the night of November 24th, 1959? Of course I do. I was sitting in the canteen eating supper with John Adams, as we had done many times that Fall. There was not a wide choice of food in those days - spaghetti or ravioli, or occasionally, fried eggs - but our thoughts were not on the meal. We had hardly spoken, our spirits were low, then John lit his pipe and said "Well, now that we have finished eating we may as well walk over and see if anything is happening". As we went in the direction of the PS buildings, I asked him "Shall we go over to the main control room or to the central building. Chris Schmelzer said that Wolfgang Schnell had got that radial phase-control thing working". John pulled on his pipe. "Probably doesn't matter, it may not do much good". Our hopes had been dashed fairly often. Then after a few more steps, he added "Let's go the central building and see what they're up to". It was about a quarter to seven.

She looked back over her time with the PS Group and was sorry that she would have to leave the next day to go back to the USA, without seeing the PS working. She continued:

> Adams opened the door of the central building. For a moment the lights blinded us, then we saw Schmelzer, Geibel and Rosset — they were smiling. Schnell walked towards us and, without a word, pulled us over to the oscilloscope. We looked... there was a broad green trace... What's the timing...why, why the beam is out to transition energy? I said it out loud — TRANSITION.
>
> Just then a voice came from the main control room. It was Hine, sounding a bit sharp (he was running himself ragged, as usual, and more frustrated than anyone) "Have you people some programme for tonight, what are you planning to do? I want to..." Schnell interrupted, "Have you looked at the beam? Go and look at the 'scope." A long silence... then, very quietly, Hereward's voice. "Are you going to try to go through transition tonight? But Schnell was already behind the racks with his Nescafe tin into which he had rapidly built his radial phase-control circuit, Geibel was out in front checking that the wires went to the right places, not the usual wrong ones. Quickly, quickly, it was ready. But the timing had to be set right. Set it at the calculated value... Look at the 'scope... yes, there is a little beam through... turn the timing knob (Schnell says I yelled this at him. I don't remember)... timing changed. little by little... the green band gets longer... no losses. It is.. look again... we're through... YES, WE'RE THROUGH TRANSITION!

The beam was stopped while the magnet power supply was adjusted to go to full excitation. Hildred continued:

> Finally, the call came through - magnet on again, pulsing to top field. Call the linac for beam... one second (time for acceleration) is a long time. The green band of beam starts across the 'scope... steadily, no losses... to transition... through it... on, on, how far will it go... on, on, IT'S ALL THE WAY Can it be? Look again at the timing... All the way... it must be 25 GeV! I'm told I screamed, the first sound, but all I remember is laughing and crying and everyone there shouting at once, pumping each others' hands, clapping each other on the back while I was hugging them all. And the beam went on, pulse after pulse.
>
> Slowly, we came back to earth. John Adams was first. Looking very calm, he went to the phone to ring up the Director-General to tell him the news but Bakker didn' t seem to grasp it right away (could it be that John was just a little incoherent?).

If John really was " just a little incoherent", then he rapidly regained his usual composure, because Sir Alec Merrison, at the memorial gathering at CERN after his death, recalled:

> I remember very clearly ... when the PS first worked and John Adams rang up to tell me. He did not tell me in highly excited tones. He said "Remember those scintillation counters you and Fidecaro put in the ring? Will they detect 20 GeV protons?" I paused long enough to grab a bottle of whisky and Professor Fidecaro, in that order, and came along to celebrate an accomplishment which I had taken no part in but was very glad to share. And that was very characteristic of John. [5]

As soon as the news got around, people began to come to congratulate the team and Giorgio Brianti remembers seeing John most disconcerted, and even angry, when Gilberto Bernadini, who was the leader of the Physics Group, came in and embraced John and kissed him on both cheeks, in a typical Italian way. During the evening, amongst other celebratory drinks, the contents of the famous vodka bottle were consumed. This story has been told many time, but must be repeated here.

A few months earlier, on a visit to the Dubna laboratory. which had been set up in the USSR as the equivalent of CERN for the Russian satellite countries, John had been given a bottle of vodka by Professor Nikitin, to be drunk when the PS exceeded 10 GeV, the energy of the world' s largest weak-focussing synchrotron being built there. At a meeting called the day after the PS had reached 25 GeV, to tell the CERN staff what had happened, John displayed the empty bottle and inserted a photograph of the

John with the famous vodka bottle and the photographic evidence of the PS reaching 25 GeV.

oscilloscope trace showing that. 25 GeV had been achieved, so that it could be sent back to Dubna .

Despite John' s efforts to make people aware of the lack of effort that was being put into providing the beam lines and experimental equipment to go round the PS, most people were taken by surprise by the rapid construction and bringing into service of the PS, about a year earlier than had originally been envisaged. Franco Bonaudi remembers that they borrowed some magnets from the Synchro-cyclotron (SC) and one from Liverpool, and these were moved around from beam A to beam B between shifts. He said "There were no power supplies. I made arrangements to rent some welding sets that ran very well with rudimentary current control".

At the following Council meeting when the successful start-up was discussed, Amaldi, who was then Chairman of the Scientific Policy Committee, said:

> This result is due to the ability and endurance of the whole PS Division and above all to John Adams, who, endowed with outstanding scientific and technical ability and human qualities, has led his colleagues since the early days of this great enterprise.

Renie had the impression that Bakker seemed more interested in the forthcoming inauguration of the PS than the deficiencies in the experimental equipment. This inauguration took place on the 5th of February 1960, attended by scientists and politicians from the Member states and visitors from the USA. Niels Bohr released a bottle of champagne to smash against a concrete block to launch the good ship PS and all who metaphorically set sail in her!

The equivalent accelerator being built in the USA, the Brookhaven 30 GeV AGS, did not come into operation until over six months after the PS reached full energy, but any advantage on the experimental front was lost by the failure to prepare adequately for the necessary experiment al equipment . Although slightly smaller than the AGS, the PS later reached 28 GeV when the magnetic field was taken up to the level where saturation effects began to affect the field distribution and the correction coils could no longer compensate.

Although the PS had operated successfully, John allowed his natural caution to show, perhaps to an excessive point, when he wrote in the first quarterly report on the operation of the P S:

> Thus the situation in December 1959 was that the synchrotron had worked successfully up to its design energy, and already beyond its

> design current, but with its builders and operators in a state of almost complete ignorance on all the details of what was happening at all stages of the acceleration process.

In fact, with improved instrumentation, understanding of the operation of the accelerator proceeded rapidly and it was not long before the design particle intensity was exceeded by a factor of ten. With additions and modifications it was later found possible to increase this by another factor of ten, and today the PS is still the centre of activity at CERN, supplying not only protons but also anti-protons, electrons, positrons and heavy ions .

Now we must go back to John's other activities in this period.

References.

1 Amaldi, E., *John Adams and his Times,* CERN 86-04. 30 May 1986.

2 Adams, J.B., Hine, M.G.N. and Lawson, J.D., "Effect of magnet inhomogeneities on the strong-focusing synchrotron," *Nature,* **171**, (1953) 926-927.

3 *Lectures on the Theory and Design of an Alternating-Gradient Proton Synchrotron,* Institute of Physics, University of Geneva, Oct. 1953.

4 *CERN Courier,* Special issue on the PS. Reprinted in 25 years of CERN, Geneva.

5 *Sir John Adams, 1920-1984: speeches made at a memorial meeting held at CERN on 27 April 1984*. CERN, Geneva, 1984.

Chapter 6: CERN POLITICS

Some of the reasons why John decided that he would have to concern himself with events outside the PS Division are best shown by extracts from the reply he wrote to a letter from Donald Fry, about a year before the PS was finished. Fry asked him if he was interested in returning to Britain to take on a new enterprise, called Zeta 2. He replied:

> Firstly, both Sir John (Cockcroft) and I have told Bakker and CERN that I shall stay to finish the PS. If I return to Harwell before finishing the machine, there will be the feeling in CERN and amongst the Member States that Britain has let CERN down.
>
> Secondly, the staff of the PS Division which I have built up during the past four years, and recently has begun to worry about the future, will take my leaving as a sign of disintegration, and this will profoundly affect the PS project. There are also my more personal relations with the senior staff of PS, some of whom have recently turned down attractive jobs outside CERN and have accepted new contracts with CERN. The confidence of this senior staff and indeed the whole PS staff would be severely shaken if I left before the machine is finished.
>
> Thirdly, there is my own peculiar position in CERN. For over two years now I have not only directed the work of the PS machine, but also been forced to take over an ever increasing part in the running of CERN and in the future of CERN. At the beginning I joined in reluctantly, just to protect the interests of the staff of the PS and to prevent any stupidities from adversely affecting the building of the machine. Later on, when Dakin left and I remained the most senior British member of staff, I found myself being drawn into all sorts of discussions of general policy from which I could hardly escape. Now with difficulties arising between the directors it is not only the senior staff that come to me for advice and encouragement, but the directors themselves. It is extremely difficult to see how I could have prevented this, except by not caring for the future and for the staff that have in many cases given up good jobs to come and work here. Perhaps I care too much about such things, but to put it at its lowest level there seems to me little point in building a large machine without making sure that it will be used. If I withdraw now, I am afraid that the shock to CERN would be as heavy as to the PS Division.
>
> Lastly, there are my own personal problems and preferences. If all the

> foregoing problems did not exist, I would jump at the chance of working on Zeta 2 because it is just the job I like and for which my few abilities are suited. Also it would solve the education problems for the children, which is by no means an unimportant one. However, I have never yet walked out of a job in the middle and morally it still seems to me a bad habit to acquire.

Apart from his loyalty, the third paragraph illustrates an important side of John's character. He was never afraid to involve himself in matters that were outside his official duties if he thought they might affect the job he had agreed to do. Bakker was a good physicist, but he had had no previous experience of directing a large organisation of the type of CERN, and was prepared to let some things go by default. John would not stand by and see decisions being taken that might be to the detriment of his group and so gradually became involved in more and more aspects of CERN as a whole. One of these was the contract system. Most of the CERN staff were engaged on three-year contracts and towards the end of the first three years some of his key staff were looking at other opportunities. It was essential to provide some incentive for them to stay on, to overcome the problems of uncertainty and difficulties of working in a foreign country, and so John pushed, against Bakker's opposition, for the establishment of the so called "indefinite" contracts. However, he could only get agreement for these to be awarded after six years of fixed term contracts, so that none could be awarded before 1959, when the PS should have been about finished anyway.

He then tried another tactic. In November 1957 he wrote to Cockcroft pointing out that the programme laid down by the Council was for the building of the two machines and their subsequent use for physics research, and the only place for those interested in applied physics and engineering was in supporting this research. Already some of his key staff were considering offers from outside CERN and if they left there would be little chance of finishing the PS by 1960. John suggested that it would help to keep some of those people if the Council, at its December meeting, were to make a statement that its intentions for the development of CERN included the study and planning of future machines. He added: "It should be made clear that there is no new financial burden implied in this statement since no project can start until after 1960 and probably not before 1962. In fact the last thing we want is any more projects before the PS is finished." The Council agreed to the setting up of a small team to look into future machines and we shall see the result of this later on in our story.

Another area where John was concerned was the preparation, or lack of it, for the experimental programme on the PS. Bakker, not alone amongst the European physicists, had little idea about the complexities of performing experiments on such a machine as the PS. It had been assumed that teams from the universities would come to CERN, complete with their apparatus and any staff needed to operate it, carry out their experiment and depart with their equipment, leaving a space for the next team. In this, so called "trucking" scenario, CERN would just run the machines and would not have to provide much in the way of support staff. A large part of the construction team for the PS would then no longer be needed and could return to their home countries. A number of people realised that this picture was unrealistic and began to call for some reorganisation to plan the experimental support. One of these was Cockcroft, who brought the matter up in Council in June 1957, apparently with little effect, since he wrote to John on 18 November 1957, "You are probably aware that we are not very happy about the way the organisation is going at the moment, particularly in respect of the organisation of work for preparing for experiments with the PS" and he wrote again in February 1958, saying that he still had no information on that subject and asking should he prod Bakker. On the experimental programme, Bakker was still convinced that there was no urgency because, as Kjell Johnsen put it, "they believed that the few pieces of equipment round the SC could somehow just be transported on a truck over to the PS on the day they had a beam". Bakker put forward a scheme for the reorganisation of the laboratory at a senior staff meeting in April 1958 and although there were repeated discussions about it, nothing was actually done. In John's reply to Cockcroft he said that there had been further meetings but nothing had been decided yet and added "Perhaps it is asking too much of the present Directors to organise CERN disinterestedly".

It was also becoming apparent to John, through discussions with experimental physicists and visits and contacts with the American laboratories, that a large support infrastructure for the experiments was needed round a machine like the PS to make full use of it, but he could not persuade Bakker to make plans for this. Other apparatus, too large for individual groups to provide, was also necessary, such as the detectors known as liquid hydrogen bubble chambers, which could show the tracks of different particles. A small bubble chamber, 30 cm in diameter, was in use at the SC and it was proposed to transfer it to the PS later on. Some people, particularly Luiz Alvarez, who had built a 72 inch bubble chamber at Berkeley in California, advocated building a 2 metre diameter one for CERN, but this was resisted strongly by Bakker and Gilberto Bernadini, who was responsible for the experimental programme of the SC.

John and Bakker at an early meeting of the CERN Council

After some argument, two transfers to the PS Group were agreed; one, a young physicist, Guy van Dardel, was to look into the requirements for the experiments and the other, Charles Peyrou, a round-cheeked gregarious Frenchman with a never-ending fund of gossip and a background of cosmic ray work using cloud chambers at the top of mountains, was to move with his team from the Scientific and Technical Services to assist in the building of a 2 metre bubble chamber at the PS. At first, Colin Ramm, an Australian who had worked on the Birmingham synchrotron before coming to CERN to oversee the magnets for the PS, was to be in charge of this, with Peyrou as his technical adviser, but this did not work out, as they could not decide on the split of responsibilities. Peyrou told me that he felt unwelcome in the Group, being allocated the less good designers in the drawing office, but it was eventually settled that he should built the 2 metre hydrogen chamber and Ramm would build a 1 metre chamber using liquid propane instead of hydrogen, this decision being endorsed by the Scientific Policy Committee in May 1958.

These two rivals later became friends, outside their direct work, and Peyrou related some advice given to him by Ramm, who said:

> John looks so persuasive when he says "shouldn't it be that way?", you believe that *he* thinks it should be that way, but that is not true at all. He's trying you. So keep to your point of view and don't allow him to persuade you. After a while he will come round to your point of view, but he will never admit that it was your point of view in the first place.

This was true of John's character; he would listen to all the proposals as to the best way of doing something; he would turn them over in his mind, probably during one of his long trips across the mountains at the week end, and then later bring forward *his* solution to the problem, which might have an uncanny resemblance to one of those that had been proposed earlier, but he would have forgotten that.

Serious work on the construction of the 2 metre bubble chamber was only started in 1958 and so it was not finished until 1962. Meanwhile, Bernard Gregory, later to be the fourth Director-General of CERN, offered the use of an 80 cm bubble chamber the Saclay Laboratory had built, and this was the mainstay for part of the physics programme in the early years of the PS. These bubble chambers were only part of the equipment required to carry out the experimental program. A big hall had to be built and magnets and other equipment needed for the beam-lines had to be ordered as well as big generators to power the bubble chamber and other magnets. Measuring machines for the analysis of the bubble chamber pictures had to be designed and built. Since there was little direct experience of carrying out experiments on this scale in Europe, it is perhaps not surprising that the physicists took some time to face up to the realities of the situation. John remarked later that instead of getting experience by joining in the experiments on the high energy American machines, the European physicists, with certain notable exceptions, had stayed at home and used the small cyclotrons in their own countries and the 600 MeV machine at CERN, which came into operation two years before the PS. He thought that in this respect the small CERN machine had done a disservice, although it had enabled research to begin quickly on a European basis and some excellent physics results had emerged. It took nearly two years to recover from this situation, since it was not until 1962 that sufficient experiments for the PS were built and put into operation and research results began to emerge. This recovery was not helped by the fact that the Council, once the construction period was over, had been led to believe that the CERN budgets could be reduced or at least stabilised at a constant level.

As the PS approached completion, John began to think about his own future. He had a fixed term contract with CERN, was on secondment from the UKAEA and expected to go back there after completion of the PS. He had turned down Fry's approach about building Zeta 2, which was just as well, as the project was never approved. He now began to make discreet enquiries as to what post he might be offered. John had thought that he might be asked to be Director of one of the UKAEA research laboratories when he had finished the PS and naturally he hoped for Harwell.

However, there had been some changes. When Harwell was transferred from the Department of Atomic Energy to the Atomic Energy Authority (UKAEA) in 1954, Cockcroft had become both Director of Harwell and Member for Research on the UKAEA Board. Although he nominally handed Harwell over his deputy Basil Schonland in 1958 he continued to behave as if he was Director even to the extent of appointing Arthur Vick to be deputy Director without consulting Schonland! He continued to serve as Member for Research on the Board until the following year when he resigned to become the first Master of Churchill College, Cambridge. Sir William Penney, later Lord Penney, who up to then had been Director of the weapons establishment at Aldermaston, was appointed to take his place. A new organisation, The National Institute for Research in Nuclear Science (NIRNS), had been set up in 1957 to supervise the whole non-secret nuclear physics programme in Britain. Part of the Harwell site was allocated to this new organization for the construction of the Rutherford Laboratory, under the direction of Gerry Pickavance, taking over some of the Harwell accelerators and equipment.

As we will see in the next chapter, it was also decided to build a laboratory to concentrate all the thermonuclear plasma research at one place. Penny thought that John was the right man to be Director of the new laboratory, to be built on a new site at Culham, not far from Harwell, and offered him the job when he had finished at CERN in 1960. John was not too happy about the way the UKAEA had been reorganised and was doubtful about accepting. Shortly before his death, Penny sent some notes to me in which he wrote:

> Schonland (then Director of the Research Group) arranged for Adams to call on me and we had a long talk. I stated as correctly as I could what the Authority was thinking about the future and the special conditions and privileges which the fusion site at Culham would have. The Director at Culham would have a free hand in the design of the buildings on the site, on the selection of new staff and on the details of the technical programme. He would be a member of the Research Group Board and would speak for Culham.

After this visit, and thinking out all the pros and cons, John decided to accept. One of the pros in this assessment was that he wanted to bring up his children in the British educational system, since they were now ranging from 8 to 14 years old. So, in August 1959, he was appointed Director Designate of the new laboratory and was expected to take up this appointment full time in 1960.

However, fate put a spanner in the works. Director–General Bakker was killed in an air crash while visiting the USA in April 1960. Amaldi was informed and made arrangements to get to Geneva as soon as possible. However, when he arrived he found that most of the necessary actions had already been carried out by John, who had taken control. An emergency meeting of the Council was called on the 3rd. of May and the Delegates seemed to have had less hesitation in appointing John to be Acting Director –General at that time than they had had earlier in appointing him to be Leader of the PS Group. His appointment was to last until a suitable replacement could be found. There seemed to be some difficulty in finding a suitable outside candidate quickly, because there followed discussions between Francois de Rose, President of the Council, and the UKAEA, to see if John could be released from his commitment to return to Britain so that he could be confirmed in the post. Penney wanted John at Culham as soon as possible, but a compromise was reached and, after another special session of Council which was called in July, it was announced that John had been appointed Director–General of CERN until the 1st of August 1961, but with the proviso that from October 1960 he would spend part of his time planning for the new laboratory at Culham.

Despite the short time he was to serve as Director–General, or perhaps because of it, John threw himself into the task of reorganizing CERN as he thought it ought to be. As we have seen before, Bakker retained the idea that CERN should be run like a university physics department, with lots of little experiments, and had not faced up the problems of running a big laboratory where things were on such a scale that good organisation and planning was necessary. Charles Peyrou went as far as to make the somewhat unkind remark that, although Bakker's death was of course unfortunate for him and his family, it was less unfortunate for CERN, because he took a very narrow view of the scope of CERN. "The Council left it to him and he did nothing. It was Weisskopf who defined the scope of CERN, not John. Weisskopf changed the stature of CERN, with the help of a very good accomplice, Mervyn Hine, who provided the technical accountancy."

Whatever he might or might not have done for the scope of CERN, John was determined to leave it with a good administrative structure. In his contribution to a book on the organisation of research establishments published in 1965, he wrote:

> The founders of CERN had conceived the laboratory as a place where the most modern facilities for high energy research would be built for the physicists of Europe to use. However it was never very clear to whom these physicists should belong. Some people argued that they should belong to the universities and should visit CERN from time to time to use the experimental facilities of the laboratory. From this came the idea of 'truck teams' coming to CERN with their apparatus to use the beams of particles from the accelerators. Other people, realising the problems of experimenting with large machines, admitted the need for some resident physicists on the CERN staff who would provide a background of experience and local knowledge of the CERN facilities. The fear was that so many physicists would join the staff at CERN that the "outsiders" at the European universities would never get time on the machines to do their experiments. The 'truck team' idea was tried out with moderate success on the SC, but when the time came to use the big machine it was realised that the visiting teams would be impotent unless integrated with CERN staff.
>
> In the early part of 1960, however, many physicists in Europe felt that CERN was not welcoming visiting physicists as much as it should, and of course the early difficulties with the experimental programme of the proton synchrotron did not allay these fears. Therefore, in addition to the reorganisation of the Divisions of CERN and the setting up of a Directorate, it was necessary to overhaul the arrangements by which external physicists could use the CERN facilities and influence the experimental programmes. [1]

To solve this problem, John set up a Nuclear Physics Research Committee (NPRC) composed of senior physicists, some from CERN, but the majority, including the chairman, from outside Universities and Institutes. The task this NPRC and its three Experiments Committees (one for each main experimental technique) was given was to examine all proposals for experiments and to determine which should be accepted and the machine time that should be allocated to them. John remarked "Much as one dislikes committees, and cumbersome as these appear on paper, the system worked remarkably well."

In the reorganisation of CERN itself, he divided the personnel of the existing six Divisions into twelve new ones, reflecting the change from construction to utilisation, and appointed four non-executive Directors, one of them being Mervyn Hine, to oversee the main areas of CERN's activities. This was not appreciated by some of the old guard. Johnsen commented:

> When John took over the PS, it had the traditional tree-like group structure. He then had the power to change it, but he didn't. When he became DG, he went away from that to this matrix system with too many people at high places that did not have specific responsibilities. It would have been better if he had copied, at a higher level, the PS structure. When he later built the SPS, he went back to the group structure, but as soon as he became DG again, he reverted to the matrix. He must have thought that one system was right for a project but the other for running a laboratory.

It was in this period, after the successful operation of the PS, that John received his first honours. The University of Geneva decided to award him an honorary DSc, but only informed him about a fortnight before the ceremony was due to take place on the 2 June 1960. Unfortunately, John had already arranged to go on a visit to several laboratories in the USSR and the arrangements could not be changed at short notice, so Renie took his place and received the accolade on his behalf. He was also awarded the Röntgen Prize by the University of Giessen in July 1960 and this time he was able to go to receive it in person. At the ceremony, he insisted that it had not been awarded to him as an individual, but as the representative of the group whose efforts had made possible the construction and early operation of the accelerator. This was followed by the award of the Duddell Medal by The Institute of Physics and The Physical Society in October of the same year, and Honorary Fellowship of the Manchester College of Science and Technology in November. Just before he left to go to Culham, he was awarded an honorary doctorate by Birmingham University in July 1961, where he was the youngest to have been given such an honour. Renie remembers that he took almost childish pleasure in this and insisted in purchasing the red Doctorial robes for the ceremony, instead of hiring them. They were subsequently used mainly for playing Father Christmas with the children. At the presentation, the Public Orator began:

> The children of Noah, when they attempted to raise themselves to the heavens by means of a tower, were frustrated by a confusion of tongues. In our day, when men are again striving to reach the heavens, it has been granted to this son of Adam to build a Tower of Babel with complete success. ... CERN is a co-operative effort in fundamental nuclear research amongst a number of European countries and today we honour the man who deserves the greatest share of the credit for bringing this scheme to fruition.

Shortly after leaving Harwell to become Master of Churchill College in Cambridge, Cockcroft wrote to John to say he was proposing to put his name up for election to the Royal Society, but warning him that "a nomination often takes several years before it becomes effective and, of course, in some cases it never does. However, I would hope that you have a good chance of election within one or two years." John was elected a Fellow in 1963, an honour that he prized almost beyond any of the others.

Meanwhile, John had asked an eminent theoretical physicist, Victor F. (Viki) Weisskopf, to come from the Massachusetts Institute of technology, where he was Professor of Physics, to become the Member of the Directorate responsible for the research Programme, a post Viki had refused when Bakker was Director–General, as he had not been very happy with Bakker's methods. He arrived in September 1960 and, as the search for the next Director–General was still going on, John persuaded him to let his name be put forward, despite his reluctance to take on the extra administrative responsibility. With John's strong recommendation, the Council appointed him Director–General, to take effect on the 1st August 1961, when John was due to return to Britain to take over his responsibilities at Culham full time. Meanwhile, it was proposed that Viki could work closely with John to learn what was involved in running an organization such as CERN. However, as the result of a car accident early in 1961, he suffered severe damage to his hip and spent some time in the Cantonal Hospital in Geneva. John did his best to help and arranged a radio link so that he could listen to what was happening in the Council and other meetings at CERN. After a while, Weisskopf decided to return to the USA for an operation on

Farewell to CERN. Left to right – Renie, Weisskopf, Eliane de Modzelewska (Secretary to the Council), John

his hip, which was successful, and he returned to CERN at the beginning of July, giving him just one month overlap with John.

Before continuing with this story of John's life, let us first take a look at what had been happening so far in the field of Thermonuclear and Plasma research.

Reference.

1 Cockcroft, Sir John (Ed), *The Organization of Research Establishments*, Cambridge University Press, 1965.

Chapter 7: NUCLEAR FUSION

The physics of fusion are complicated, but the main thing to concern us here is that, if a plasma of ionized nuclei of the lighter elements could be confined and heated to a sufficiently high temperature, the nuclei would fuse together with a release of energy, as in the case of the sun, whose temperature is maintained by this mechanism. If such a reaction could be carried out in a controlled manner, it could form a source of energy for the future that is potentially limitless and could be free from many of the dangers associated with fission power from nuclear reactors.

Although the possibility of obtaining energy from nuclear fusion had been suggested just after the first world war and subsequent experiments had demonstrated fusion between isotopes of hydrogen nuclei, it was not until after the second world war that serious work started to find out whether controlled fusion could form a useful source of power. Because of the possibility of producing intense sources of neutrons and their application in the weapons field, information concerning this process was classified about 1950, although it was suspected that work was going on in the USA, Britain and the USSR. This veil of secrecy was lifted a little in April 1956, when a state visit to Britain was paid by the Russian leaders Kruschev and Bulganin. Accompanying them was the distinguished physicist Igor Kurchatov, who suggested to Cockcroft that the Harwell people might be interested in having a lecture on the thermonuclear reactions in gas discharges, describing the Russian work in that field. This took everyone by surprise, but the offer was gladly accepted. After this Russian initiative, it was inevitable that the policy of secrecy in this field of work should be reconsidered, and some relaxation was soon evident. Early the following year, the British Physical Society published a group of papers, including one by John Lawson, the Harwell accelerator physicist, whom we have come across earlier as the voice of doom for the strong-focussing principle. He was now involved in this work and had calculated the conditions necessary, but not sufficient, for a sustained reaction: the Lawson criteria as they came to be known.

Other scientific publications began to appear, and it turned out that the work going on in the three areas, although carried out independently, tended to be along the same lines. In order to keep a plasma at high temperature, millions of degrees, without it being cooled by the walls of the containment vessel, meant that it had to be held away from the walls by means of magnetic fields. The problem was to find a configuration of magnetic fields that would do this in a stable fashion. There were many different approaches to this problem and one that had been studied in all three areas used a toroidal vacuum chamber, with windings to give both transverse and longitudinal magnetic fields inside it.

A major part of the British effort in this field was carried out at AERE Harwell where a very large, for its day, magnetic confinement device was designed in 1955 and came into operation in August 1957. This was called Zeta (Zero Energy Test Assembly) since it was intended to demonstrate the principle, without producing any net gain of energy. The Americans wanted to delay any publications in this field until the second "Atoms for Peace" conference to be held in Geneva in September 1958, where they planned to put on a big exhibition to offset the technical advantage the Russians had obtained by the launch of the first "sputnik". However, the British Establishment wanted something to take the public heat off the Windscale fire, and announced the results of the first tests of Zeta at a press conference held at Harwell in January 1958. This attracted great publicity, as neutrons had been detected, which some people thought indicated that fusion had taken place in the plasma in Zeta. Even Cockcroft, relaxing his usual caution when flattered by a lady journalist, was cajoled into saying 'yes' to a question whether he was 90% certain that thermo-nuclear reactions had been achieved, a statement he was to regret later. [1] However, it was subsequently demonstrated that the neutrons were the result of some collisions between particles accelerated in the fields produced by the plasma. The press, having blown up the first results into virtually free power for all, now ran Zeta down as a failure. It had in fact achieved everything for which it was originally designed. It was modified and useful work on gaining a better understanding of plasmas was carried out with Zeta right up to the end of 1968, when it was shut down.

In 1956, long before Zeta was finished, Harwell were already thinking about a bigger machine, Zeta 2, and a preliminary design was worked out. There was sufficient optimism about the prospects of this proposal that D.W. Fry wrote a letter to John, on 24 March 1958, which included the paragraph:

> We want to build Zeta 2 now and in fact several people are already thinking about it. If there was any possibility of our interesting you in taking control of it I am sure this would be warmly welcomed at

Harwell. Would you be willing to entertain the idea or not?

Extracts from John's reply to this have already been given in the previous chapter. He was certainly tempted, but considered his duty to others first. It was a field that required the same techniques as used for accelerators and the same kind of components, such as magnets, vacuum and radio-frequency systems and large power supplies. John already had some knowledge of what was going on in that field, since he had contact with the work going on at Harwell in the years before he left to go to CERN. Evidence of this is given in a talk given by Bob Carruthers in 1988, in which he recalled: "With a small capacitor bank, a pulse transformer borrowed from John Adam's klystron development project and a torus put together from some handy chemical glassware bends, we produced restricted discharges lasting up to 100 microseconds". [1] With John's quest for knowledge and desire to master every subject he came in contact with, it is certain that he would not have made this loan without investigating the work for which it was to be used.

In the event, Zeta 2 was never built. The design evolved and it became too large and expensive, aiming at plasma currents that were only achieved many years later by the JET machine built at the Culham site as a joint European venture, and Zeta 2 was abandoned in February 1959. Meanwhile, other work on plasma physics was going on at the Atomic Weapons Research Establishment (AWRE) at Aldermaston and at the Associated Electrical Industries (AEI) laboratories near by.

Before the abandonment of Zeta 2, Cockcroft had decided that he wanted a reactor test site, as he was convinced that all the different design possibilities should be explored before deciding on the type of reactor to choose for the nuclear power stations that should be built in Britain. This was strongly opposed by Christopher Hinton, who was Member for Engineering and Production and was responsible for the design and construction of the power stations. After a furious row, Cockcroft managed to persuade the UKAEA Board to support him and, in the autumn of 1858, he chose a site at Winfrith, in Dorset, for the new laboratory and put Fry in charge of it. He also wanted to move the Harwell plasma work to Winfrith as additional justification for setting up the new laboratory. When this proposal was announced, the move was strongly opposed by those concerned, but the main reason for rejecting it, according to Penney, was the desire to concentrate most of the fusion work, including a big new experiment which was being planned, in one place, free from the security restrictions of Harwell and Winfrith. A suitable site for the new laboratory was found at Culham, not far from Harwell, and it was purchased from the Admiralty in May 1959. The new experiment was called ICSE (Intermediate Current Stability Experiment) as it was considered to be a stepping-stone to the next stage in producing power from fusion.

Earlier, in March 1958, a number of scientists got together at CERN to talk about plasma physics experiments, mostly the results from Zeta, as these had just been published. As a result of this and the general freeing of information as to what was going on in this field, CERN was asked by a number of laboratories to extend this informal study group and the June Council formally approved the setting up of a CERN Study Group on Fusion Problems, to "exchange information, discuss the programmes being undertaken in various laboratories and to consider ways of facilitating fusion research in Europe", with John as chairman. The first meeting of this study group was held in September, shortly after the second "Atoms for Peace" conference had taken place in Geneva. At this conference, the results of research in this field were revealed by papers presented mainly by workers in the USA, the USSR and Britain, the three areas where a substantial amount of work was going on, and the Americans mounted an impressive exhibition. The first meeting of the CERN study group was attended by about 50 scientists from 10 European States, and it became clear that so much new information had been presented at the conference that it would take time to assimilate it.

On taking on anything, John always tried to get it quite clear what the objective was and, after this December meeting, he wrote in his notebook:

Objects of work

The avowed object is to make power from the fusion reactions in the following stages:-

a) Heat a plasma to ignition temperature in a container,
b) Having reached fusion temperature, to maintain reacting plasma in a container,
c) To get more energy from the reacting plasma than is being put in to maintain the temperature and contain the plasma,
d) To get out of the device the excess energy in a convenient form.

He added, with characteristic realism:

> For this result many countries are prepared to spend millions of dollars. It is very important to see this connection if money is to be avail-

> able for fusion work, i.e. if the object was only plasma physics, only money on on academic scale would be provided. Under the umbrella of fusion work plasma physics may prosper, but it is only of secondary importance from the point of view of getting money.

He then went on:

> Not even stage a) above has been reached yet. The main reason why this is so is lack of physical understanding of the behaviour of plasma in any environment. The technology needed for the experiments is adequate ... but can be improved. Measuring techniques are in general not adequate for the experiments. Either the experiments must be made simpler or the techniques of measuring plasma greatly improved.
>
> Thus the problem has been attacked with vast resources of money and technology but with little physical basis and the first attack has not succeeded in reaching the first objective.

He examined the results obtained so far and came to the conclusion that large complicated projects should temporarily abandoned in favour of simple idealised systems which might or might not be on a large scale. For large scale experiments he suggested building up multi-purpose "facilities". He ended his note with the examination of a number of different possible experimental projects and the kind of laboratory and staff that would be required to carry them out.

A number of people have expressed surprise that John, at this period when he was fully occupied in the latter stages of the PS construction, could find time to acquire enough knowledge about this subject, of which he had previously only had peripheral contact, to gain the respect of some of those who were experts in the field. One can see, from the above extract, that, even at this early stage in the Study Group's deliberations, he had analysed the situation and come to his own conclusions. The report, published in April 1959,[2] recognised that the big machines had demonstrated some of the principles, but had disappointed some of their sponsors, because basic physics understanding had not advanced sufficiently. Smaller scale experiments to try to understand this basic physics were necessary.

The Study Group considered whether a European laboratory was needed but concluded that "unless it can be demonstrated that one is needed in order to build larger facilities that can be built by national groups, the many other advantages of such a centre may prove insufficient to overcome the difficulties in its creation and maintenance." Since the CERN constitution did not cover the long term support of fusion research, the report proposed the formation of a "European Society for Controlled Thermonuclear Research" and proposed a statute for such an organisation.

In 1985, when the John Adams Lecture Theatre at Culham was dedicated, R.S. (Bas) Pease recalled this period. He said:

> Undoubtedly many people were thinking on the same lines as Adams's study group, most notably the Euratom authorities. Nonetheless the 1959 study group report carries all the hallmarks of Adams's skill: a thoughtful and comprehensive analysis of a complex problem involving physics, engineering and what might be called statesmanship. It is this multi-faceted capability that made him such an outstanding leader in European science. Adams's 1959 report was, in the event, followed by the setting up by Euratom of the European fusion programme which indeed fostered a diverse but planned coordination of national programmes and then somewhat later by the setting up of the plasma physics division of the European Physical Society in the 1960s.[1]

John had included some representatives of the newly formed Euratom Consortium in his study and he went so far as to propose that any such research could be carried out in a laboratory adjacent to CERN, as he was always on the lookout for other activities to carry on at CERN if the funding for high energy physics should tend to dry up, but this idea encountered strong opposition from some countries and was not pursued further.

As we have seen earlier, the timescale for John's return to Britain was upset by Bakker's death, leading to the CERN Council pressing for a delay, so that he could act as Director-General until a suitable replacement could be found. The UKAEA agreed that John could stay on until August 1961 as long as he could work part time at Culham after August 1960. It is amazing that John could get so much done in this period. We have already seen what he was doing at CERN and at the same time he had also to plan a completely new laboratory. In this he received invaluable help from Denis Willson, who had been Secretary at Harwell and now moved to take the same position in the new laboratory, and Bas Pease, a larger-than-life character with a strident laugh, who had taken an important part in the earlier plasma work at Harwell.

Willson wrote:

'At Culham he faced three main tasks:
- To bring together the fusion research teams from Harwell and Aldermaston into a single unit.
- To ensure that Culham conducted research on plasma confinement physics on a wide front, covering many different lines of magnetic confinement and diagnostic development, while at the same time continuing research on the Zeta experiment (which remained on the Harwell site).
- To ensure that Culham's research was conducted with the maximum of international collaboration. [3]

The last point emphasised that all the research there was to be free from any security classification, and to implement this John introduced a scheme whereby up to one quarter of Culham's research scientists should come from overseas laboratories for periods of up to three years and then move on elsewhere.

Because of his delayed arrival at Culham, John did not have quite as much influence on the architecture as he would have wished. The first stage, which included the building for the ICSE experiment, as well as some offices and laboratories, had already been designed and approved, with the work starting in May 1960. For the second stage, planned to start in August 1961, there were arguments between those who wanted interconnected laboratories for freedom of communication, and those who wanted their separated little empires, as at Harwell, where separation had been necessary to contain possible radioactive contamination. John was in time to influence this decision and came down on the side of the interconnected layout, as he considered that communication between different groups was important. Bas Pease said that John took the layouts as they had been done and moulded them to his purpose. In the execution of this plan, Willson expressed the opinion:

> In the construction of the Culham Laboratory, Adams achieved a partnership between the architects, the civil engineers and the Laboratory Scientists which could serve as a model for anyone else planning the layout of a new laboratory. He said to me "For the moment we are planning a lab. for nuclear fusion research, but who knows what might happen? I think the lab. we build would take as an example the National Physical Laboratory, or the research department of a large electrical manufacturing company — anything that uses a similar blend of research and supporting services, with plenty of electric power at

high voltages and precision workshops.

The Culham site had been an airfield, with a triangle of runways which were wider than the roads needed by the Laboratory. The surplus width was turned into car parks, so that from the outset there was parking for 500 cars, so we did not suffer from the problem so many

Luncheon discussions in one of the courtyards at the Culham laboratory.

> laboratories have. To assist effective collaboration between the research teams coming from different sites, separate buildings were avoided as far as possible: John Adams' plan was for a long central building containing services that everyone would use, with "wings" extending from it containing experimental halls and small offices. To get the right blend of laboratories and services, some of these wings were extended until they enclosed small courtyards in which grass, shrubs and trees were planted. [3]

The central building included a spectacular lecture theatre, with a fully cantilevered roof, which was named the John Adams Hall after his death. An example of John's attention to detail is shown by his insistence that the ash-trays should have rubber rims, so that the pipe smokers could tap their pipes out without disturbing the seminar.

While he was still at CERN, John, together with Bas Pease and Denis Willson, produced the Culham Laboratory First Planning Report in May 1961. This report, not for general publication, surveyed the possible ways of achieving power through thermonuclear reactions and the existing work being carried out in the various laboratories in the UK, ending up by giving the proposed transfer of staff from other establishments and estimates of the additional staff and money required for the proposed programme. The idea of a planning report was new to the Authority, but John's idea was that if it was accepted it gave him the permission to carry it out.

Even before this, Penney was becoming worried about the ICSE experiment. The physics of plasma confinement by magnetic fields was clearly much more complicated than the simple theory supposed, and the cost estimates, which he considered to be of the "black of an envelope" type, were already up to £3M, a large sum at that time, and still rising. In the notes he sent to me, he wrote:

> Adams had not been involved in the choice of ICSE as the major experiment to be put at Culham. I thought it possible that after six months he would say that ICSE is alright but what he really wanted to do was something else. My conclusion was that the choice of ICSE should be challenged and that a better suggestion was that Culham's task for the first few years should be to obtain a better understanding of the physics of the plasma.
>
> I invited Adams and Pease to my house one evening after work and put my thoughts to them. Adams said he was glad that I had spoken. He was not entirely happy with the ICSE proposals and he liked my sug-

> gestion of Culham's first task to understand better the physics of plasma. However, did my suggestion mean that the Culham budget would be severely limited, to which I replied no. Pease was then asked for his opinion. He was clearly a little disappointed but he also saw the advantages of the new charter, so he gave his support.
>
> It was agreed that Adams would explain the decisions to the fusion staff and that I would ask the Authority to approve the cancellation of ICSE.

John's aim was to build up a firm basis for future developments, not to get stuck on one line of approach but to provide the flexibility to switch to others as progress was made. In his December 1958 notebook he had remarked that there are a large number of items, such as generators and condenser banks, as well as diagnostic equipment, common to many of the different types of experiments. He now followed this up in the definition of what he called "Experimental Assemblies": collections of equipment likely to be useful for a number of experiments. This would not only avoid waste but would also give a new experiment a flying start. By the time John gave up his post as Director of Culham in 1967, 17 of these experimental assemblies had been built and operated and three more were under construction. From this it can be seen that the fusion programme had moved from trying to make a break-through at one point to a general advance across a broad front.

Denis Willson had found a house for the Adams at Boars Hill, which was about half way between Culham and Oxford, and they moved there in August 1961. This was convenient for the family, as the eldest daughter, Josephine, went to the High School in Oxford, Kate was already at St. Helen's in Abingdon and Chris went to a preparatory school in Oxford.

From the start, John believed that the universities should continue to be involved in the plasma work, as they had been in the beginning, since he was sure that could be to the advantage of both parties and he lost no opportunity of encouraging such contracts. He had a number of friends at Oxford University and one of these was Nicholas Kurti, who was not only Professor of Physics but also an expert in the culinary arts, having written a book on the physics of cooking, a most unusual combination! When he wrote to Renie on the occasion of John's death, he recalled that:

> As soon as Culham was established I felt strongly that a laboratory devoted to unclassified research should be associated in some form with the University of Oxford and I mentioned to John the idea of

> establishing a Professorship of Theoretical Plasma Physics. Unlike an experimental professor who would rightly regard Culham as his home a theoretician would be mainly working in Oxford while maintaining close contacts with Culham. John liked the idea, so at my suggestion the Principal of Brasenose — John, as you know, was a member of our Senior Common Room — arranged a luncheon for a small group of University scientists and administrators to discuss this possibility John explained, that while the UKAEA could not fund such a professorship, he as the Director of Culham might be able to fund it for an initial period of 7 years. A week or so later John had a meeting with the Registrar who proposed as the next step, the drafting of an agreement between Culham and the University. Whereupon John reached into his pocket and presented the Registrar with the draft contract. This must have made a great impression on the Registrar because things from then on moved speedily — in fact embarrassingly so. Thus some of the Colleges heard about the proposed Chair before their physics tutors had an inkling and barely 5 months after the proposal had been made. Congregation passed the decree on 12 October 1962.

Culham provided financial support for this Chair, which was entitled "Mathematics (Plasma Physics)", for the first two or three years and then the University took it over until 1989, when it was announced that the Chair would be allowed to lapse.

Another contact with the University was through Isiah Berlin, a Fellow of All Soul's College, who was helping to raise additional funds for a new graduate college to be called after Lord Wolfson, since his fund was providing the major contribution to the cost. Knowing John's passion for architecture, Berlin invited him to join the building committee. When the college was finished, it was proposed that John should become a Fellow, in view of the contribution he had made to its design. Since one of the requirements for this was that he should be a university graduate, he was awarded an MA degree in record time and without examination!

The main contribution that John made to the fusion programme was in the organisation of the Laboratory, the assessments of the various programmes to be undertaken and the steps taken to try to ensure that once a programme had been decided, the means to carry it out were made available. He wanted to know and understand the scientific reasons and technical problems of each experiment, but did not have the close involvement in their design as he had had in his earlier work on the Harwell cyclotron and the CERN PS. As Lawson put it, he had to make judgements on what other people said and to decide who to back of the various different factions that were inside Culham, without a very deep knowledge of the plasma physics. Amongst these factions were Peter Thoneman, who had taken a prominent part in the earlier work at Harwell and who thought he was in the running to become Director of Culham before John was chosen, who was an advocate of one type of confinement scheme, and Brian Taylor, of whom John had a very high opinion and who was an advocate of another type. It was a difficult balance to maintain.

The Laboratory was organised into eight divisions under the control of the Culham Management Committee, consisting of John and the heads of the divisions. Project teams were formed from physicists and engineers to construct the experiments. Bas Pease recorded:

> Adams was extremely anxious that the project officers and departmental division heads should take total responsibility for each project, financial, staff, design, construction and operation. This reflected the CERN requirements and practice, rather than the previous Harwell practice, where the finance was, as it were, laid on together with the electricity and other site services. He instigated the Culham Annual Report, as a means of describing to the Authority and workers through-

The Phoenix II experiment and the team which included two Russian scientists.

> out the world the progress being made ... These reports provide the most complete record available of events at Culham.[1]

I do not propose to go into the various experiments that were carried out during John's period as Director of Culham, except to mention one, Phoenix II, in which scientists from Russia were involved for the first time. John reviewed the whole field in the 1965 Guthrie Lecture given to the Physical Society[4]. Stafford wrote:

> The scientific achievements at Culham during the Directorship of John Adams were substantial and bear his hallmark of careful and painstaking development towards the understanding of a very complicated physical process.[5]

Not long after he returned to Britain, John received his first recognition from the British Government for his work at CERN. It is not often that Britons are honoured for their work abroad, unless they are in the Diplomatic Service, which is presumably why he was made a Companion of the Order of St. Michael and St. George (CMG), an honour mainly reserved for those in that Service. This time Renie managed to persuade John to hire the necessary morning suit to go to Buckingham Palace to receive the award, but complained that "after the ceremony I imagined that he would stop somewhere for a celebratory lunch. Not so. He wasn't going to be seen in any pub or restaurant in this 'get up', and so I had to drive home with him absolutely starving".

Before continuing with the rest of John's stewardship of Culham, we must see what had happened as a result of one of his other activities.

References.

1 *Plasma Physics and Controlled Fusion.* **28**, 397-419.

2 Adams, J. B., *European Fusion Research; report of the CERN Study Group on fusion problems.* CERN, Geneva, (CERN 59-16)

3 Willson, D., *A European Experiment*, Adam Hilger, Bristol, 1981.

4 *Proceedings of the Physical Society*, **89**, 189-216, 1966.

5 Stafford, G. H., John Bertram Adams, 1920 – 1984, *Biographical Memoirs of Fellows of the Royal Society*, **32** (1986) 3-34.

Chapter 8: INTERLUDE AT THE MINISTRY

Soon after he returned to Culham, John got increasingly worried about the way that science and technology were being organised and funded in Britain. He had a number of friends and acquaintances in high places who might be able to influence the politicians and he began to write to them, giving his views as to how things should be organised. He tended to get polite acknowledgements, but two of the recipients of such a letter must have taken more notice. These were Professor P.M.S. Blackett (later Lord Blackett), a Nobel Prize winner, who had played an important part in the discussions over the creation of CERN, and Vivian Bowden (later Lord Bowden), an old friend from the TRE days who was then Principal of the Manchester College of Science and Technology which had awarded John an honorary Fellowship in the middle of 1960. Both these friends happened to be closely associated with the Labour Party.

When the first Wilson Labour Government was elected in 1964, Bowden was made Minister of State for the Department of Education and Science. He approached John to see if he would take an active role in solving some of the problems, but Renie wrote:

> John had never had much time for party politics and was not interested in joining the Labour Party. He had no particular allegiance to any political party and always voted for the party candidate he felt would be best at a particular time. He could be classed as a member of the "floating voter". First and foremost he was a builder, whether it was a laboratory or an accelerator. He liked to plan the intricate details, putting them together, as he once said, like a jigsaw puzzle until he had eliminated the greatest difficulties.
>
> John did have views on how science and technology in Britain should be organized and had been horrified by the lack of interest from British Industry when contracts of about 100 million Swiss francs were being placed for the building of the PS. While the rest of Europe was competing for these contracts Britain stood aside.
>
> He thought that if we could plan our research and development resources in the many government laboratories to collaborate more closely with industry, it would be of great economic benefit to the country. Lack of planning of our research and development resources in government labs on the one hand, and our inability to use the some-

> times brilliant ideas they engendered to the benefit of our industries on the other, meant that countries other than Britain were taking advantage of our research facilities at our expense.
>
> He thought that a great deal needed to be done in the training of engineers and improving their status. Many of the University Engineering departments were too theoretical and divorced from the real world and in some of the university science faculties "industry" was almost a dirty word. There was no engineering school at all at the University of Oxford. The "brain drain" was in full flood. Young physicists and engineers were going over to the USA where they were appreciated and where there were many more opportunities. We were in need of a more prestigious training like that given at M.I.T. in the USA or the Technische Hochschule in Germany and Switzerland. He was interested in the new Colleges of Advanced Technology known as CATS where he hoped that there would be more direct contact with industry perhaps in the form of sandwich courses where students would spend some time working on the factory floor. He also believed that it would be in Britain's long term interests to join the EEC.

To try to implement his battle cry that he was going to produce the "white heat of technological revolution", Wilson set up a Ministry of Technology to deal with the problems of Britain's industry. He realized that without the good will of the trade unions it would stand no chance of success, so he chose one of the most prominent trade unionists, Frank Cousins, Secretary-General of the Transport and General Workers Union (TGWU), to lead the new Ministry. Blackett, although not officially part of the new government, was retained as scientific and technical adviser, while keeping his position as Professor of Physics at Imperial College. Charles Snow, distinguished scientist and novelist, famous for his "two-cultures", accepted a peerage and became Parliamentary Secretary in the House of Lords. Maurice Dean, a mathematician and an experienced civil servant, with only two years to go before retirement, was appointed Permanent Secretary.

As time went by, Frank Cousins was spending more time arguing with the Prime Minister about the need for an income policy than in reorganising the technological base of industry, and finally resigned over this issue. In addition to this, the major organisation for which the Ministry was responsible was the Atomic Energy Authority, which had a large military programme, and both Cousins and Snow were well known to be opposed

A cartoonist's idea of Cousins and Snow at the Ministry of Technology.

to nuclear weapons, and so could be put in an embarrassing situation at any time by a question on this subject. Neither had shown much ability to make a good speech or to debate well in Parliament, or to frame a clear political directive in the Ministry, which was a relatively small spending department of the Government. Lord Penney wrote: "Whitehall did not so much oppose the new Ministry and its people; it ignored them". Cousins was replaced by Tony Benn, an extreme left-winger who sometimes let his enthusiasm for a cause overcome his critical faculties.

According to Penney, Blackett had become worried and even desperate. He was telling everyone who would listen that the future well-being of British industry depended on having a strong computer industry, both hardware and software. He decided that the Ministry needed a high-grade technologist. He knew Adams and thought he was the man. Penney's notes related:

> In the traditional method of British science, Blackett invited me to lunch at the Athenaeum (in the snack bar, because neither of us wanted a full lunch) and put the proposition to me. I thought he had a good case. Would it be necessary for Adams to be full time at the Ministry?

> Blackett thought not – at least not immediately. What the Ministry wanted was someone with ideas, who would suggest actions which could be taken by others. It was suggested that John would spend half time, or perhaps a little more, at the Ministry and the rest at Culham. I thought this could be managed.

Blackett urged John to take up this offer and to put into motion some of the things he had been proposing earlier. John agreed to take the job of one of the two Controllers in the Ministry, but only for a trial period of one year. He would remain Director of Culham and share his time between there and Whitehall. He took along with him two members of the Culham staff, Ken Binning and Vernon Birchall.

In a contribution John made to the biographical memoir written by Bernard Lovell for the Royal Society on Blackett's death, John describes the situation he found when he took up his position in Whitehall.[1] He wrote:

> The Ministry of Technology when it was first set up consisted of two parts and the relationship between them was by no means clear. There was a small number of civil servants assembled in London under Maurice Dean, The Permanent Secretary, to deal with the industries sponsored by the Ministry and a relatively large number of people in the R & D (Research and Development) establishments in various parts of Britain which the Ministry had inherited from the DSIR (Department of Scientific and Industrial Research).
>
> Soon after I joined the Ministry, two branches were set up at Millbank Tower, one for the establishments and the other for the industry and Ieuan Maddock joined me to run the industries branch. At this time Mintech was very small and it had developed a good pioneering spirit.
>
> One of my main preoccupations was how to use the people in the R & D establishments to help with the industrial problems. Blackett on the other hand took very little interest in the establishments and seemed to regard them as irrelevant for the purposes of the Ministry which were to improve the performance of British industry. Many ways of interacting with industry were developed during this period, some more effective than others.

John went on to say that, although Blackett was passionately concerned with these industrial problems he had very little first hand knowledge of industry. He had been very successful during the war using Operational Research methods to combat the U-boats but these methods were difficult to apply to the Government-Industry relationship. He pointed out that industry could not be commanded like an army, nor was it the enemy of government, so the analyses and strategies of war situations were hardly applicable. Nevertheless, he believed that Blackett hoped to develop something analogous to operational research methods which could be applied to the industrial problems in Britain. He ended up:

> During the 18 months I was associated with the Mintech I cannot claim that any coherent method was developed but many ways of interacting with industry were explored and in these Blackett played an energetic role.

John must have created a bit of a disturbance when he took up his office at Millbank Tower, as the first thing he asked for was a blackboard. He never worked without a blackboard in his office, to illustrate his ideas to colleagues and visitors. Here, it was most important to have one to explain to others his ideas on such things as matrix management, where the organisations supplying services formed the vertical columns and projects needing the services of several organisations formed the horizontal rows. His reorganisation of CERN had embodied some of these principles, but now he developed it further. No blackboard could be found, so John made a telephone call to Culham, one was dismounted from an office wall, put on a truck and delivered to his new office.

The Ministry provided him with a chauffeur-driven car for his transport between Culham and Whitehall. So as not to waste the travelling time, John designed a folding desk and desk light and had it fitted to the car, as well as a radio-telephone, a rarity in those days. This was done for practical purposes, but it was also one of the ways that, perhaps unconsciously, that John managed to get himself known to people outside his immediate field. He was known to more than one member of the Cabinet as 'the man with a desk in his car' and was even featured on the front page of *The Sun* newspaper, with a photograph of him sitting at his desk. [2]

Soon after starting this work, John became convinced that it would be very difficult to mobilise the very considerable resources available to government for the support of British industry under the existing system, with the scientific and technical establishments reporting to different departments of the government. He proposed that a new organisation should be set up, outside the normal departmental system, to analyse the needs of industries and the value to them of technologies to be developed, and to draw up programmes for the research establishments to work towards satisfying these needs and ensure that technical developments made in these establishments were exploited commercially. This proposed new organisation was given the name "British Technology Authority"(BTA).

As time went by, and his proposals fell upon deaf ears, John got more and more frustrated. It seems that his difficulties were not all with the establishments as Tony Benn reported in his memoirs that on one occasion John Adams, the Controller, seemed quite unable to articulate what he thought should be done and it was only afterwards that he learned that John had put in a paper on this very subject and the Permanent Secretary was sitting on it and didn't want the matter discussed.

Penney related how John would come to his office to "let his hair down". He said there was a great chasm between British industry and the Ministry, which had research and development establishments whose purpose was to help industry, but John was far from convinced that they were as effective as they could be. Penny had already helped him by providing a deputy Controller to be responsible for creating a structure for communication between the Ministry and industry. This was Ieuan Maddock, mentioned above, who had "worked himself out of a job" with the Weapons Group, "a good and fast talker on any subject, and an electronics expert with many contacts in industry". John had taken to him immediately and had offered him the job. He later became a key member of the Ministry.

In May, Penney, who was by then Chairman of the UKAEA, was proposing that John should be made Member for Research on the Board, replacing Arthur Vick, who was resigning to become Vice-Chancellor of Queen's University, Belfast. He wrote "I would ... like an announcement made about you as soon as possible, but I would be quite willing for you to spend half your time (or even a little more; I leave that to you) in the Ministry of Technology from the date of your appointment (which I would like to be the 1st. October)."

However, John was not to spend much more time at the Ministry, as his frustration got the better of him and expressed this in a letter he wrote to Sir Maurice Dean in June 1966. [3] After the introduction and some proposals for the future, John stated that the job which was outstanding and which interested him was the "Controller of Programmes", namely the job of forming and evaluating new programmes for all the Ministry of Technology and Ministry of Aircraft stations and the AEA. He went on to detail what this would involve. He followed this by a statement that he would be interested in this work but could only do it under conditions which would ensure its success. These he stated as follows:

> So long as the AEA exists as a separate statutory body I would have to have an inside position in the AEA at Board level; otherwise there is no effective way of exerting a direct influence on the programmes of the AEA and the redeployment of such places as Harwell and Aldermaston. Bill Penney has asked me to be Research Member of the AEA Board and that would suffice providing that non-atomic programmes come under the Research Member.
>
> Secondly, the "Programme Group" would have to be located near the centre of gravity of the large stations whose programmes are the concern of the Group and whose staffs will be involved in the Group's work .... (which) is in the Oxford area. This is in easy reach of London (1 hour by train) and London Airport (1 hour by car). It is also near Oxford, Reading, Surrey and London Universities, many of whose staff I would aim to use. For example, Nuffield College and the Centre for Management studies at Oxford could both be very useful allies. The Group should be located at Culham, which is an attractive locality from the point of view of recruiting staff and very close to Oxford. It has good library and computing facilities and other necessary amenities and these facilities are backed by many others at NPL, Harwell, Aldermaston and Oxford.
>
> Lastly, I must have the freedom to attract high calibre staff to the Group and this means flexibility in salaries, secondment conditions and consultancies ...
>
> This seems to me about the only way I can help in the future. In March of this year I produced a scheme for the management of the Mintech R & D resources and an interim arrangement to get on with the programmes work. Both schemes have been discussed *ad nauseam* without any decision and without any alternatives being suggested and this despite all the hard work we have put in over the last six months.
>
> To put in crudely, Maurice, you hired me as a world expert on research laboratories, R & D programmes and advanced technology. I have done my stuff, analysed the problem and suggested a solution which other experts, such as Bill Penney, agree is the only feasible one. I have been very patient with objections of non-experts and other interested parties and tried to answer their objections sensibly. I have even learned to play Whitehall cricket in an amateur way. You have supported me whole-heartedly and our Minister says he agrees with our plans. All this has been done at the expense of the Culham Laboratory, the AEA

> and my own scientific career. Apart from setting fire to myself in the middle of Whitehall I cannot see what more I can do.
>
> I agreed to join you at Mintech for one year which terminates at the end of July. Unless there is something else substantial I can do in the Ministry, and I do not see anything at the moment apart from keeping the paper moving and cheering up the troops, I propose to return to Culham at the agreed date.

By putting so many conditions on trying to ensure the success of his proposal, John virtually made it certain that it would not be taken up. However, he had not entirely given up. In a letter to Sir Richard Clarke, the then Permanent Secretary, in July , he wrote

> If I became Research Member of the AEA, I could also run the Programmes Group, but then there would be nobody of sufficient scientific stature left in the Ministry to forward the BTA scheme and to manage the Research Stations and Establishments.

He went on to point out that if the BTA scheme did not go ahead, he would opt for the AEA part, since he was convinced that he could organize that part successfully . He followed this by a letter in which he wrote:

> You are most kind to refer in your letter to the contribution I have made in setting up this Ministry. I count it a great privilege to have been allowed to take such a full part in the affairs of Whitehall and to have been accepted so wholeheartedly as a colleague by senior civil servants. I have formed my personal attachments and I have learnt a great deal. I would certainly like to remain in close collaboration with you and the Ministry in the future, but rather than become a consultant or adviser of some sort I would prefer doing a definite job for the Ministry such as the one I have mentioned above.

In the formal letter appointing John as the Member for Research, Tony Benn wrote:

> As you know, I shall be very sorry to lose your direct help in the Ministry. I am most conscious of your major contribution in your year here to what we are all trying to do. I look forward, however, to your joining the Board of the Authority where I know your qualities will be great asset; and I am glad that we will continue to have the benefit of your advice as a member of the Advisory Council on Technology and also the more direct contribution which you will still be making to the Ministry's work.

Despite all these polite sentiments, the question has been asked why John's efforts seemed to have made so little impression at the Ministry compared with his success on the European scene. When questioned later, Binning thought that John tried to do too much too quickly:

> Being director of a laboratory with a great deal of power to act in one's hands is totally different from the way things happen in governing a country. There is an enormous inertia to overcome. Most people think the ideas he outlined were and could be made to work given time, but he was only at Mintech for twelve months and that only half time. Being in London only in the afternoons made it difficult for him to meet busy people. He could be very persuasive and had already made many people see that the scheme was necessary.

Another reason may have been that he was working in a totally different system. At the Ministry he was working with both politicians and civil servants in a system where the politicians make the decisions and the civil servants carry them out. The politicians tend to be transient and are usually not experts in the subject and so have to rely heavily on the civil servants for information and advice on technical matters. Since the civil service provides the continuity in a transient political situation, it resists change. Therefore, any proposal for reorganisation is likely to be delayed by setting up committees of all interested parties, with sub-committees to look into individual aspects, giving infinite scope for delays by referring back from one committee to another and so on. Unless it is one of the Minister's pet schemes, it may never be seen again. The situation at CERN was quite different. The vital decisions were made by the Council, which was made up of two delegates from each Member State, irrespective of size. Normally, one delegate was a senior scientist, representing his country's physicists and the other a senior administrator representing the government. The important thing was that these were delegates with the right to commit their country to the decisions taken by Council. Of course they had to consult their colleagues at home beforehand, represent their views, and get any necessary financial authority. The decisions made by Council could be executed at once; they did not have to be ratified by a Council of Ministers or a European Parliament, as now is the case for most subsequent joint European ventures. John was much more at home in this sort of environment. He was arguing with experts in both fields, the scientific and the administrative, people who, if they were convinced, would go back home to try to influence those concerned to agree to that particular proposal. The politicians at that time were mostly in favour of European co-operation: a successful European laboratory like CERN was a good symbol, as long as it did not cost too much.

Years later, John took enormous pleasure in watching the television series "Yes, Minister". He said it exactly mirrored his experience of life in a Ministry.

References

1 Lovell, B., P.M.S. Blackett, *Biographical Memoirs of Fellows of the Royal Society*, (1976).

2 *The Sun* newspaper, Dec. 19th. 1968. Unfortunately, the paper was unable to supply a copy of the photograph to be reproduced here.

3 Stafford, G.H., John Bertram Adams, 1920 – 1984. *Biographical Memoirs of Fellows of the Royal Society*, **32** (1986) 3-34.

Chapter 9: MEMBER FOR RESEARCH

When Penney had agreed to release John to work part-time at the ministry it was because, as he recalled, "For a few years, Culham thrived and there were no scares, either technical or financial." However this happy situation on the financial side was not to last. By the time John returned to work full time for the Authority, the Labour Government's hopes for the success of its policies were turning sour, with cold winds blowing on the financial front. The balance of payments was getting worse and the International Monetary Fund was putting conditions on any further support. The Minister for Technology, Tony Benn, called for a review of the UKAEA programmes and, as usual in such circumstances, it was the long term research that had to suffer. Fusion was no exception. Although Culham was then only costing £4M a year, a small fraction of the total expenditure of the Authority, a special "Review Panel" was set up, which included members of the Authority and experts from universities and industry. Denis Willson, the Culham secretary, wrote:

> It is not possible to report the deliberations of this review panel, which were never published, but its recommendations to the UKAEA were clear and drastic.
>
> After examining the weight of evidence presented to them, the panel concluded that nuclear fusion had some potential as a new source of energy, but only in the long term. No-practical impact upon the British economy could be expected for at least twenty years. In the economic climate of 1967, the panel did not think that this justified the scale or the cost of the Culham Laboratory, and they recommended that the Culham fusion research programme to be cut in half. The UKAEA accepted the recommendation, but felt that to avoid unnecessary hardship, the cut should be spread over five years. In other words, a 10% cut should be made each year in the staff and cost of fusion research at Culham, until by 1972 the desired result had been achieved.
>
> This recommendation naturally came as a shock not only to the Culham Laboratory, but also to fusion research workers elsewhere. Nevertheless it was approved by the Minister of Technology and put into effect. There was some expectation in the UKAEA that this hard-headed appraisal of nuclear fusion research might lead to a similar action by other countries, but that did not happen. Fusion research else-

> where went on expanding so that, despite the pioneering work in Britain..., the British share of the world effort fell drastically.[1]

Some people think that the recommendations of the panel were influenced by the resentment that John had stirred up in his attempts to introduce reforms when working for the Ministry, although no evidence has been found for this. Bob Carruthers, who acted as secretary of the Panel recalls that there was an equal split between those who wanted to carry on the programme as it was and those who wanted to close it down entirely, the resulting cutting by half being a compromise!

John took part in this Review Panel as both Director of Culham and as a member of the Board and one can guess how unhappy he felt to be party to the cutting down of the budget for the laboratory he had built up with such pride only a few years earlier.

Since the fusion budget was to be cut by half, Culham was left with the choice of reducing the staff of the laboratory or finding contract work which was paid for out of other budgets. The latter choice was taken and John managed to get a portion of the UK space programme transfered to Culham. In this diversification, John was helped enormously by Bas Pease, who was taking an increasing part in the running of the laboratory and introduced a fighting spirit to go after other work to avoid any redundancies.

In addition to Culham, Harwell was also feeling the cold economic wind and the necessity to slim down and look for outside contract work. John was able to use his extensive contacts to assist in this endeavour. At the same time, he continued to press for the co-ordination he had advocated at the Ministry, this time with some success, being instrumental in the setting up in February 1967 of the Programmes Analysis Unit (PAU), funded jointly by the Ministry of Technology and the Atomic Energy Authority, to analyse the research and development programmes over the whole field covered by these two organisations. This success may have been helped by the fact that John was a member of the Council for Scientific Policy from 1965 to 1968.

With this and other activities, John took less and less direct part in the running of Culham and handed this over to Bas Pease, who was made Head of the laboratory, but not Director until about a year later.

Since his departure from CERN in 1961, John had kept in close touch with what was happening in the high energy physics field. This was aided by several official appointments. One was to the NIRNS Board in February 1961, where he could exert some influence over the British programme. On the European scene, he was a member of the CERN Scientific Policy Committee (SPC) from 1961 and acted as its vice-chairman in the years 1964 to 1968, which kept him in close touch with all that was happening there. When Cockcroft retired as British delegate to the CERN Council at the end of 1961, Lord Hailsham, then Minister for Science, asked John to take his place for an initial period of two years, which was subsequently extended for a further year. At this time, Britain was rather negative to increase in the CERN budget and, as we shall see in the next chapter, took a strong stand on the priorities for building new machines, and John had to convey this attitude to the Council. His unhappiness about his conflicting loyalties is shown in a letter he wrote to Bannier, then President of the Council, in April 1965. In it he wrote:

> I am sorry that I shall not be present at future Council meetings when all these decisions are made. On the other hand, the somewhat bitter irony of having to question the development of one's own brain children will now cease and that makes parting easier for me.

Returning to the fusion programme, John had always been aware that there was a big gap between achieving "ignition", the situation when a plasma becomes self-sustaining, and the building of a practical reactor as a source of useful power, which involved enormous technological problems that had not so far been studied in any depth. He now set up a group to look into these problems, which produced a report in 1967 [2] concluding that a fusion reactor was possible and, subject to developing knowledge of the physics of the plasma, it could be made an economical proposition. This was the first detailed examination of this problem in Europe. However John was not blind to the social and economic problems involved, since he wrote in his report to the Fusion Review Panel in 1981:

> It was also claimed at the outset that nuclear fusion was a "clean" energy source, compared with nuclear fission. However, the neutrons released ... activate the structure of the reactor and maintenance and repair operations will produce a certain amount of radioactive waste. Also, tritium is a radioactive isotope and must be well contained ... to a level acceptable to the general public. Finally, like any other energy source, fusion reactors will produce waste heat but, differing from coal or oil burning power stations, no chemical waste.
>
> Nuclear fusion reactors will not, therefore, be entirely safe nor wholly clean. They will have their potential dangers and environmental risks

> but compared with other major energy sources they have definite advantages in these respects which makes them attractive.

However, John knew that the optimism of the scientists could clash with the pessimism of the economists and at one meeting he quoted:

> A very successful and very wise economist John Maynard Keynes wrote in his book on General Economic Theory: If human nature felt no temptation to take a chance, no satisfaction (profit apart) in constructing a factory, a railway, a mine or a farm, there might not be much investment merely as a result of cold calculation.

Although his position as Member for Research made him responsible to the Board for the research programme of both Harwell and Culham, it did not give him much executive power, because he was not made Director of the Research Group. Since Walter (later Sir Walter) Marshall had taken over at Harwell and neither he nor Pease at Culham wanted any interference with their plans, John was left with less to do than he would have liked. Never one to be idle, John sublimated his earlier frustration by writing a number of papers giving his ideas on the role of science in industry and on the part that government could play.

About this time John was wondering what his future would be. He only had a five year contract with the Authority and he had some hopes that he might get the job of Chairman of the Authority and have his contract extended. These hopes were dashed when John (later Sir John) Hill was appointed to this post. Therefore, when he was approached in April 1968 about the possibility of his going back to CERN to build a new accelerator, he said he was interested, being already a member of Amaldi's Steering Group that was discussing this new machine and so he was aware of all that it entailed. So when the formal offer was made the following December, he accepted willingly, despite the uncertainty whether the accelerator would be authorised. His heart was really more in constructing big projects than in the wheeling and dealing in politics, unless the latter was in aid of progress in the former. Politics, however, showed in the letter he wrote to the Deputy Chairman of the Authority on deciding to take up this offer, in which he wrote:

> You can well imagine that this decision has not been an easy one for me. I am much attached to the Authority and to the two laboratories of the Research Group and it has been a great privilege to have been able

to play a leading role in establishing our new laboratory at Culham as a world centre for nuclear fusion research and to help in restoring Harwell to its old vigour with new programmes and a dynamic management. I am particularly happy about the way the PAU has developed since such a body, independent of the laboratories, seems to me essential in establishing a proper R & D investment policy.

It was clear also in a letter to Benn, which included:

Of course there are great risks attached to this job and there are many problems yet to be resolved but the challenge is so great and the goodwill so strong both in this country and in the rest of Europe that I feel I must accept the risks with the job and hope for the best. My main regret, as you know, is that our Government did not find it possible to join in this second programme of CERN and thus make available to it the technical and managerial wisdom of this country which was so important in ensuring the success of the first European programme at CERN, Meyrin. I can only hope that the reasons for this decision will change in the future and that before long the U.K. Government will be able to review its position in more favourable circumstances.

As we shall see, his wish was to be granted. Despite this decision to go back to CERN, John continued to retain interest in, and make contributions to, the field of fusion research. One of his last acts before leaving Culham was to give strong support to the proposal for a joint experiment by a Culham team on the Russian T-3 tokomak, a toroidal confinement configuration that was by then accepted by many people as the most likely to succeed in producing the conditions required. However, Stafford remarked:

It is perhaps surprising that during his time at Culham, Adams did not take more positive steps towards securing his vision of a European programme for fusion by promoting stronger British association with the European Group. This was to some extent rectified many years later when in 1974 he accepted an invitation to join the Joint European Torus (JET) Scientific and Technical Committee, which worked during the design phase after the essential framework of British participation in the Euratom programme had been settled. There can be no doubt that his support for JET was of great importance in securing approval for the project. When a Director came to be appointed to the JET project, Hans-Otto Wüster, who worked with Adams during his second period at CERN, was appointed. [3]

John's association with this new European project went further than joining the committee and supplying it with a Director. In 1976, he had been approached to see if he would take over the direction of JET himself, but could not accept as he had only recently become Executive Director-General of CERN. Over the following six months there were discussions about sitting JET in France, close to CERN. John encouraged the idea, pointing out that CERN could then provide the majority of the professional services, or could even contract to build the complete machine, thus reducing the difficulty of building up a large staff for the construction, which would later have to be reduced when it was finished. Nothing came of this, and a site next to John's old laboratory at Culham was chosen, so he still had at least emotional connection with the project.

The last major contribution to fusion made by John was some years later, as a member of a panel set up in 1981 to review the European Fusion Programme. This was known as the Beckurts Review Panel after its chairman, who was a Research Director at Siemens, Munich. John took an active part and drafted much of the report, which endorsed the Commission's main recommendation for strategy and programme for 1982-86. He was preparing to serve on the second Beckurts Review Panel in 1984 when he became ill. He was also chairman of an International panel in 1983 set up to assess a proposal to build a large thermonuclear experiment called IGNITOR, which aimed to push technology to the extreme to reduce the size and cost of a reactor. The problems were assessed to be too great and the project was abandoned.

With his acceptance of the position at CERN, John had once more to uproot himself and Renie from a comfortable English country life to a rented house in Geneva. However, this time the children's education did not play any great part in the decision to move. Josi had graduated in physics and was already working at CERN, being somewhat annoyed when she heard that John was to return there, as she wanted to make her own career, without the shadow of her father over her. Kate had taken an Arts Diploma and was teaching in London and Chris was at the Atlantic College, so for the first time for years they were on their own. Once the project was approved, CERN bought a house for the Director-General at Founex, in the Canton of Vaud, with an option for John to buy it later, which he duly exercised.

Before continuing with this story, we must turn the clock back once again to look at the events that led up to his return to CERN.

References.

1 Willson, D., *European Experiment: The Launching of JET*, A. Hilger, 1981.

2 Carruthers, R., Davenport, P. A., and Mitchell, J.T.D., *Culham Laboratory Report*, CLM-R 85.

3 Stafford, G. H., "John Bertram Adams, 1920-1984." *Biographical Memoirs of Fellows of the Royal Society*, **32** (1986) P3-34.

Chapter 10: THE 300 GEV PROJECT

As we have seen earlier, even while the CERN PS was still in the middle of its construction phase, the accelerator builders were already beginning to think of the next stage. The 1957 June Council meeting was told by John of the proposal to set up a small group in the PS Division to carry out research on new ideas and the December Council approved this, stating that "the studies of the group, under the leadership of Mr. Adams on new ideas for accelerators are of greatest importance to CERN and to the Member States", but warned that this did not commit then to any great expense or new project.

At first, the main effort of the group, consisting mostly of visitors to CERN, under the guidance of Arnold Schoch, who had come from Heidelberg University in the early days of the PS, was concentrated on the plasma betatron, one of a number of new ideas for accelerators coming from the USSR. They soon came to the conclusion that this was unlikely to be of practical use in the near future and turned their attention to the possibilities of colliding one beam of particles with another. When a high energy particle, such as a proton, hits a fixed target only a portion of its energy can go into producing a required event, but if two particles moving with equal energy in opposite directions collide, the combined energy of the two particles is released. The idea was to have two rings of magnets, intersecting at one point, and to accumulate beams of protons from the PS, rotating in opposite directions in the two rings. The group, which subsequently became the Accelerator Research (AR) Division, was strengthened enormously at the end of 1959, by the transfer to it of the key personnel involved in the construction of the PS, which had then come into operation. Instead of being a group looking at long term possibilities, without any great urgency, it was now invaded by those who had just built a successful machine and were eager to go on the next one. They concentrated on the storage rings and produced a first design towards the end of 1960 and, early in 1961, it was decided to build a model, using 2 MeV electrons to simulate 25 GeV protons, to check the possibility of "stacking" a number of injected pulses into a storage ring to increase the current in the circulating beam.

Meanwhile, a number of other European laboratories were building, or planning to build, smaller accelerators of different kinds and, on the initiative of two of these laboratories, a 'European Accelerator Study Group' was set up and John was asked to chair the first meeting, held near Paris at the Saclay laboratory in May 1961. In his opening address, John said that the creation of the Group had been stimulated by the fact that the armoury of accelerators in Europe was entering a new phase; in planning and designing the next generation of accelerators, a reasonable amount of coordination would help to avoid too much overlapping by allowing the new equipment to be complementary rather than competitive. He also reported on the design of the storage ring proposal for CERN, which now had concentric rings, to give eight interaction points for experiments, and had been given the name ISR (Intersecting Storage Rings).

When the idea of the ISR as the next machine for CERN began to be proposed early in 1961, there was a strong reaction from the physicists. There had been talks in the USA about higher energy machines, in the 200 GeV region or even higher, and these seemed more attractive (but also much more expensive!). One of the strongest opponents of the ISR was Leon van Hove, then head of the Theory Division at CERN, who called a meeting of nuclear physicists in June to find out what they wanted. At the following meeting of the SPC, John reported a lack of interest for storage rings and the emphasis now seemed to be on a high intensity proton accelerator of about 100 GeV.

During 1962, the nuclear physicists started talking about the needs for higher energy machines. To put the discussions on an organized basis, Viki Weisskopf, who had taken over from John as Director-General of CERN, and C.F. Powell, famous for his cosmic ray work at Bristol University and then chairman of the Scientific Policy Committee, convened a meeting of physicists from CERN and the European countries, in January 1963, to consider the desirable characteristics of a future accelerator and the means for exploiting it experimentally. This meeting, in which John took part, decided to form the European Committee for Future Accelerators (ECFA) and appointed Amaldi to form a working party to study the requirements. The members of the working party met nine times between January and June, when they produced a report [1] giving their conclusions. These came down strongly in favour of a programme which should include the construction of both a pair of storage rings, the ISR, for the CERN PS, and a new proton accelerator of very high energy, about 300 GeV, and high intensity. It also recommended the building of smaller machines in the individual member states, to form "the base of the pyramid". A later ECFA meeting set a target of 10 to the power of 13 protons per second for this machine, ten times the intensity that the PS had achieved at that time.

A feasibility study for a proton synchrotron and a laboratory to satisfy these aims was started in 1963 at CERN by a team under Kjell Johnsen and "The Design Study of a 300 GeV Proton Synchotron" was published November 1964 [2]. Although both the ECFA and CERN studies considered building the new machine close to the existing CERN laboratory, it was concluded that no site available in that vicinity was large enough. Therefore it was proposed that it should be built elsewhere in Europe, on a site in one of the Member States.

In a paper presented to the 8th. International Conference on High-Energy Accelerators, held in Geneva in 1971, John wrote:

> In view of the subsequent events, I should also mention another important step which took place in the autumn of 1964. Several of us in Europe, realizing the high cost of a 300 GeV machine and a new laboratory and having experienced the difficulties of financing the previous 28 GeV machine, had the idea of basing the 300 GeV construction on a wider collaboration of participating countries. Why not, we argued, build the next accelerator as a joint project between the United States of America, the Soviet Union and the CERN Member States. A meeting was consequently arranged to discuss this idea in Vienna in the autumn of 1964 and was attended by scientific representatives of these areas of the world. At the time of this meeting, both the USA and the CERN Member States were considering new machines in the 200-400 GeV energy range and the Soviet Union was constructing the 70 GeV machine at Serpukhov. The conclusion of the meeting was that the machines up to about 500 GeV should continue to be built on a continental basis but that for machines above the 1000 GeV level intercontinental collaboration would probably be necessary and should be explored. [3]

This was the start of a campaign for international collaboration that John pursued for the rest of his life.

A proposal was put to the CERN Council in June 1965 to build a 300 GeV machine and a new laboratory as part of a package of new European facilities, the other parts of which were the addition of Intersecting Storage Rings (ISR) to the CERN PS and the addition of a booster synchrotron to the latter to increase the intensity of the proton beam. At its meeting in December 1965, the Council approved the second and third parts of this package, but came to no decision on the 300 GeV program. This was despite strong opposition from the British delegation, of which John was a member. While John strongly supported the construction of the ISR from the technical point of view, he joined with the other members of the delegation, who wanted an all-or-nothing situation at that moment, fearing that if the ISR was approved, the financial situation would lead to an indefinite delay in the 300 GeV project. This stand, which went so far as the British delegation saying they would not support the ISR, was fully backed up by the British physics community. This was true also of the other European physicists, including some of those at CERN, who preferred a higher energy PS to the ISR. However, some governments thought that by approving the ISR they might postpone for some time the additional financial demands of the 300 GeV programme; just what the physicists feared.

The story of the ISR does not belong here, as John had no part in its construction, except for the legacy he left behind at CERN in the pursuit for excellence that he instilled in the team who built the PS, the greater number of whom were now involved in the design and construction of the ISR. Although, in the opinions of some, the results from colliding two 28 GeV (later 31 GeV) proton beams together were somewhat disappointing from the point of view of the experiments that were carried out, which missed some of the discoveries which were possible at the collision energy achieved, there is no doubt that the construction of the machine, under the able direction of Kjell Johnsen, was a technical *tour de force*. In Mervyn Hine's words:

> The work that was done on that machine in those five years changed the whole way people thought about accelerators; colliders became possible; people began to understand the beams in real detail. It was an incredibly important piece of work that was done on machine physics.

Meanwhile, in the Unites States, a design study for a 200 GeV machine went ahead at Berkeley in California, and a report was produced in January 1965. At the end of 1966, after competition from both the East and the West for the site of the new machine, a compromise was made by the choice of Batavia, near Chicago, for the establishment of a new "National Accelerator Laboratory" (NAL) to build the machine and a Director was appointed early in 1967. This was R. R. Wilson, a brilliant but somewhat idiosyncratic physicist who had built an electron synchrotron at Cornell University. He transformed the project into a 200-400 GeV machine with many changes, including the use of the separated-function focussing scheme described later.

In Europe, there was rather a lull in the 300 GeV machine design, but great activity in other matters relevant to the 300 GeV Programme. On one hand, ECFA was active in the machine utilisation studies, producing a four-volume report in May 1967, which indicated that there would be no shortage of experimental proposals, and on the other, the search for sites for the new laboratory went apace and a "Steering Group", including John, was formed to look into other aspects. Most of the Member States offered one or more sites, the original 100 being whittled down to 22 by the application of some simple criteria. Even 22 was too great a number for the CERN team to evaluate, so more comprehensive criteria were laid down and the countries that had offered more than one site were asked to choose which was their premier one. Rivalry between three different regions in Germany and two in Italy could not be resolved, so there were still twelve to choose from. The CERN team, which at times included John as a member of the Scientific Policy Committee, toured the sites, being fêted and bombarded with facts as to why that particular site should be chosen. By December 1967, the number had been cut down to five, with one each in Germany and Italy, and a "Site Evaluation Panel" was set up to compare not only the suitability of a site from the technical point of view, but also what attractions it would have for the staff and their families. To pave the way for the project, the Council agreed, also in December 1967, on the changes needed in the CERN constitution to allow it to cover more than one laboratory.

In Britain, under a Labour Government which put a greater emphasis on applied rather than fundamental research, any official enthusiasm for the project was further reduced by a contentious report that, even if a British site was chosen, the net financial advantage to Britain would be negative. As a result of this, Brian Flowers, who was then Chairman of the Science Research Council and the British delegate to the CERN Council, had the unhappy task of announcing, at the meeting in June 1968, that under the then existing financial situation Britain could no longer agree to join in the 300 GeV project, but he expressed his personal hope, and that of his colleagues in Britain, that the project would go ahead.

The Council had decided that the choice of site would not be taken until after sufficient states had agreed to join the programme, to avoid any 'site bargaining' and, towards the middle of 1969 it seemed that a decision was imminent, as six Member States had indicated their intention of joining the 300 GeV Programme. Anticipating this, Bernard Gregory, then Director-General of CERN, had already asked John if he would come out to CERN to plan the construction of the new machine and laboratory. As we have already seen, he had little hesitation in accepting, despite the uncertainty, and the June Council appointed him as project leader and Director-General designate of the new laboratory. However, support from six out of the eleven Member States was considered insufficient to approve the project, the cost of which had already been scaled down from nearly 1800 million to just over 1400 million Swiss Francs, but it was hoped that other States might be persuaded to join the Programme, so that a decision could be taken at the December meeting.

John's paper to the International Conference in 1971 included:

> Looking back, I think one can discern a number of reasons why our Member States hesitated to reach a decision on the 300 GeV Programme in the form it was presented at that time.
>
> In the first place the economic situation in 1969 for science in general and nuclear physics in particular was very different from the ebullient years around 1964 and 1965 when the 300 GeV Programme was first put forward. It was evident that several Member States of CERN and possibly all of them found the cost of the Programme too high compared with their other investments in science and with the growth rates in their total science investments, which had dropped from figures around 15% per annum in the 1965s to a few percent per annum in 1969.
>
> In the second place, the idea of constructing a second European laboratory for nuclear physics remote from the existing one, which had seemed attractive in 1965, looked inappropriate in 1969, particularly since it implied running down the existing CERN laboratory when the new one got under way.
>
> In the third place, so many delays have occurred in the 300 GeV Programme and the American machine was coming along so fast that an eight year Programme to reach experimental exploitation seemed too long.
>
> Fourthly, it turned out that choosing one site amongst five technically possible sites presented non-trivial political problems for the Member States of CERN. [3]

What John did not report was that, just before the December Council meeting, the German delegate created a shock wave by announcing that, contrary to the agreement previously reached by the Council to decouple the choice of site from the decision to participate, Germany would agree to pay its full contribution to the cost of the project only if its site, at Drensteinfurt, was chosen. As Germany was the largest contributor to the finances of the organisation, this looked like the end of the project.

Going back to when John returned to CERN early in 1969, he at once started to look into the earlier design for the 300 GeV machine, in the light of subsequent advances in that field. During this early period of his return to CERN, he acted in an uncharacteristic way, going around saying, even in Council, that he was doing the redesign with one man and a secretary. Even as late as August 1970, he wrote, in answer to an invitation to give a paper at the 1971 American Accelerator Conference, that he could not manage this because "as you probably know, there are only two of us working full time on the CERN 300 GeV project, Ted Wilson and myself." This was literally true, but very misleading, because there was a lot of part-time effort going into the design, both inside and outside CERN, as we shall see.

On his return to CERN, John poses in front of the 300 GeV design he inherited.

It is difficult to explain this behaviour. He certainly felt, whether it was true or not, that he was considered by Kjell Johnsen, who had led the team which had produced the 300 GeV design so far, as an interloper. Kjell was just coming to the end of the construction of the ISR and was the senior accelerator expert at CERN who, but for John, might have been chosen to lead the new project. John probably felt he had to assert himself, to distance himself from the previous designs, which had been criticised by some as being out of date and ignoring the new developments at NAL and elsewhere. Therefore it is probable that in his mind the 300 GeV project was the new study he was carrying out and he wanted to show that he was not being influenced by the earlier design. Although E.J.N. (Ted) Wilson, a physicist who had come from the Rutherford Laboratory to work on the project, was his only full time assistant, there were also others working on the project almost full time, as well part time activities of others, such as the Survey Group assessing the various sites. In addition, John had started a "300 GeV Machine Committee" in June 1969, made up of members from CERN and the major European laboratories, to look into every aspect of the design. It is significant that, in the letter asking him to become a member of this Committee, John starts "Dear Professor Johnsen" instead of "Dear Kjell"! Johnsen said:

> We resented this attitude very much. I never understood why he did it, because he gained so little and at the same time made many people unhappy at not getting the slightest recognition for helping.

Although I had previously made his acquaintance at various meetings, my own first long-term contact with John was when he started this 300 GeV Machine Committee. I was then a member of the Directorate of the Daresbury Nuclear Physics Laboratory, in Cheshire, England, which I had been invited to join when it started in order to help in the building of the 6 GeV electron synchrotron NINA. This was mainly as a result of my earlier work on linear accelerators. The members of the committee were asked to form a number of study groups, to look into the different components of the proposed machine. Since most of the machine was well covered by acknowledged experts in each field and there was less interest in the control problems, John asked me to act as convenor of the group looking into that subject, as I had been concerned with controls, amongst other things, at Daresbury.

As the result of the work carried out by this Machine Committee, a report was issued in the middle of 1970 [4], written by John and Ted Wilson, in which a number of new possibilities were explored. The main one of these concerned the magnet system.

Most earlier strong-focussing machines had used magnets with curved pole-pieces to produce both the dipole guiding field and the quadrupole focussing field, and so were called "combined-function" machines. Since the maximum field in an iron-cored magnet is limited to about 2 Tesla by saturation effects, the central guide field could not be greater than about 1.2 Tesla. For the 200-400 GeV machine to be built at NAL, R.R Wilson had proposed the use of separate dipoles and quadrupoles, showing that the increased field possible in the dipoles more than made up for their reduced length, needed to accommodate the separate quadrupoles. Thus, for a given circumference, a "separated-function" machine could go to a higher energy. This configuration had been studied at CERN during a design review carried out early in 1969 and the conclusion had been reached that, for the same energy, a combined-function machine would have a circumference about 15% larger than the equivalent separated-function machine, but that the costs would be about the same, so there seemed to be no reason to change from a well-tried system.

Ted Wilson had been an advocate of the separated-function scheme and had proposed such a design earlier, but had been told "not to rock the boat". John soon realised that one of its main advantages was the flexibility it gave in the design of the accelerator, rather than just the ability to obtain a higher energy from a given size of machine, as in the case for the NAL design. The idea was to build a machine which could be tailored to cover two possible scenarios. Due to the long time-scale of eight years for the programme, there was some pressure from part of the physics community for a lower energy machine that could be available earlier. By installing all the separate quadrupoles, but only part of the full complement of dipole bending magnets, the machine could be operated at a lower energy, say 200 GeV, somewhat earlier. This became known as the "missing magnet" machine. Later on, the remaining dipoles could be installed, the machine realigned, and full energy attained. The other need for flexibility was to counter a difficulty John saw looming on the horizon; that there could be a call to delay the construction to take advantage of recent developments in the technique of making superconducting magnets.

Magnets wound with superconducting cable, which has negligible resistance to a steady current at temperatures a few degrees above absolute zero, were already being used in some big detectors. For some time attention had been paid to the possibility of making air-cored superconducting magnets for accelerators, which might provide useful fields greater, by factors of two to three, than the conventional iron-cored type. This would enable much higher energies to be obtained from a given size of machine. However, for synchrotrons, where the magnetic filed has to change during the acceleration cycle, the losses are not negligible and they can raise the temperature of the cable to a point when it is no longer superconducting, known as quenching. The Rutherford laboratory, which had been formed at Harwell to take over all the non-secret particle physics work, was carrying out research on this problem, searching for the best method of making a low-loss conductor. The most promising line seemed to be to use a composite composed of superconducting filaments in a metallic matrix, which they thought might be possible to manufacture with only a moderate extension of existing capability.

The Rutherford team were successful in developing such a cable and started to design a superconducting magnet suitable for use in an accelerator and some people were already pressing for the design of the new machine to be changed to incorporate superconducting magnets. However, John realised that there were many hurdles to be cleared before magnets suitable for this purpose could be demonstrated and, even when that had been achieved, a long development period would be required to be able to produce hundreds of them with sufficiently reproducible characteristics. By adopting the missing-magnet scheme, John kept the options open. To start with, only half the conventional iron dipoles could be ordered and the superconductivity work at the Rutherford Laboratory and elsewhere encouraged. If, when the first set of dipoles were nearing the end of production, the superconducting magnets had reached a suitable stage, the machine could run initially with the half set of dipoles and the superconducting ones would be installed in the spaces later. Further into the future, the iron dipoles could be replaced by a second set of superconducting ones, to give an energy approaching 1000 GeV (1 TeV). Alternatively, if the new development did not progress fast enough, the second set of iron dipoles could be ordered and installed. It was still assumed that the machine would be built on a new site away from CERN.

Another change was in the size of the vacuum chamber inside the magnets. In the design of the PS, the size of the vacuum chamber had been chosen to accommodate the size of the beam, plus the sum of the expected beam deviations from the centre line due to a number of causes. Experience had since shown that these deviations could be reduced to much smaller amounts by various corrective devices, so that most of this "safety factor" was no longer essential. This proved to be very fortunate for the PS, as it could then handle the much more intense and bigger beams that were needed for the later developments. To reduce the size of the vacuum chamber for the 300 GeV, and thus the size and cost of the magnets, it was proposed to make it just big enough to get a "pencil beam", with its deviations, round the machine initially, relying on the ability to reduce the deviations sufficiently during the running-in process to allow a full size beam to circulate. John welcomed this idea, as it enabled the cost of the magnet system to be reduced significantly, to quieten the critics who compared the previous CERN estimates with those for the American machine.

These changes had been strongly opposed by some who had been members of John's old PS team and arguments both inside and outside the Machine Committee became heated. John would just listen and let everyone have their say. Very often in the ensuing discussion, into which John would drop the odd word if he thought it was going in the wrong direction, there would evolve what he thought was the best solution to the particular problem. He would then step in, summarise it and pass on the next problem.

When it seemed that the project was doomed after the German ultimatum, it must have struck hardest on John, who had given up a secure position in Britain to take over the leadership of the new laboratory. However, he was not one to give up easily and always seemed at his best when things looked black and there was a problem to solve. At once, he began to look for any possibility of making modifications that might save the project. The two main difficulties seemed to be the cost and the choice of site. Both these difficulties might be overcome if the new machine could be built near the existing CERN, but that solution had been rejected earlier on, partly on the basis of the site being unsuitable, but mainly due to the fear by some delegates and physicists that CERN was already too big and the power too centralised and this would become worse unless there was a new laboratory to counterbalance it. Some borings had been carried out to the north and east of CERN as early as 1961, instigated by Colin Ramm, who had been a strong advocate for a site nearby, and who had made a model to show to the Council how a somewhat smaller machine could be fitted in.

When Kjell Johnsen was asked whether his team had perhaps not searched too hard for a suitable CERN site, in case a 300 GeV at CERN might have interfered with the approval of the ISR, he denied it, saying:

> I don't think we were afraid for competition with the ISR. If it was the case, it was unconscious. We were looking for sites round this area, and had carried out a bit of a survey when we were told by the President of Council, and some Council members, that we should stop considering a CERN site. When Bannier (the President) says "Geneva is dead, don't bother to think about it", and he says it in Dutch, you take that as an instruction. We didn't resist that, because some of us believed the advantage of building up a new place, when we could see the old one would go stale.

There seems to be some controversy as who first put forward the idea of re-examining the possibility of fitting the 300 GeV machine into the CERN site. Ted Wilson is sure that John was thinking of it right from the start, since he seemed to take little interest in the design of the injector synchrotron, which would have been needed on a site other than CERN, where there was the possibility of the PS being used as the injector. Once the festive season was over, although not so festive that year for John, he threw himself into examining the area on the French side of CERN. One of the conditions previously put on the site choice was that there should be sufficient area of flat ground. This was because all previous accelerator tunnels had been constructed by digging out a trench, building the machine tunnel at the bottom of it and then putting the earth back on top. Although the possibility of boring a tunnel had been considered, the best possible site was being sought and so a condition had been placed in the specification that it should be flat enough for this normal type of construction. However, if a machine of the size proposed was to be sited near CERN, this "cut and fill" method could not be used; it would be necessary to bore an underground tunnel. Local geological maps showed that the land between Geneva and the Jura mountains was made up of a layer of moraine of varying thickness over a bed of molasse, a sedimentary rock made up of a mixture of sandstone and marl. Moraine, a glacial deposit of stones and sand, is not good for tunnelling in, but molasse is ideal, being not too hard and virtually impervious to moisture. Although some borings had been done at the time of the earlier considerations, the depth of the top of the molasse was not known with any precision over a large area, so more information would be needed before John could be certain that a machine of this size could be fitted in the good rock.

Apart from solving the site problem, assuming the French would agree, putting the new machine near CERN would open the possibility of considerable financial savings, not only by using the existing PS as a proton injector instead of building a new one, but also by using some of the existing laboratory facilities, which would have had to be duplicated on another site. It could also allow the time scale for the first experimental utilisation to be reduced, by sending beams to one of the existing experimental halls, and would not require the running-down of CERN, thus meeting all four of the objections that John estimated had led to the failure of the original proposal. It is interesting to recall that it was then estimated that the staff requirements would be reduced from the previous total of 7400 for the two separate laboratories to 5000 if they were combined. In fact, the programme was completed within a total complement of just over 3500 by the use of temporary contract labour.

So far, these ideas had been kept to the very few people involved, as it was not clear what the reaction of the Member States and the European physics community would be. Ted Wilson remembers John wondering whether he could get hold of the data from the earlier borings without letting Jean Gervaise, the head of the Survey Group, into the secret, but finally said "I suppose I must".

The result of the earlier borings showed that it should be possible to bore a tunnel round a ring 1.8 kilometres in diameter entirely in the molasse, if it was sufficiently far below the surface. This depth turned out to be under 20 metres where the ground level was lowest, rising to over 60 metres at the high point. Part of the ring would be in Switzerland, but the greater part would be under French soil. The 1.8 km diameter was quite a bit smaller than the 2.4 km of the previous design, but would still allow the machine to go up to 300 GeV with the separated-function magnet system. Ted Wilson had broken his leg skiing and John, nothing daunted, had dragged him out of hospital to work out a suitable arrangement of magnets (known as a lattice) to suit this new diameter.

Once this was done, John and Gregory made discreet approaches to a number of governments, and especially Germany, to sound out their reactions. When they proved to be on the whole favourable, the next step was for John to explain the new scheme to the March meeting of the CERN Committee of Council, a restricted meeting of the Council, who liked the idea, especially as it reduced the estimated cost of the project, but wanted

The Committee of Council meeting in which the new design was revealed. Left to right – John, Gregory (Director-General), Amaldi (Council President), Hampton (Director of Administration).

the opinion of the Scientific Policy Committee before agreeing that it should studied further. John had written a confidential paper, dated 6th. April, comparing the previous proposal, which he called 'Scheme A' and the new one, 'Scheme B', but he went on a visit to the USA and left it to Gregory to present it to the SPC and to reveal it to the rest of CERN, as he knew it would cause a furore, especially amongst the physicists, and hoped this would die down a little by the time he returned. The SPC recommended that money should be made available for more borings and design studies in the second half of 1970.

A week-end meeting of ECFA was called in early May to explain the new Proposal. As expected, this turned out to be a stormy session, in which some delegates expressed dismay at the idea of the second laboratory being abandoned and the Geneva one being enlarged. The reduction in diameter of the machine was strongly criticised, since then it would be smaller than the American machine, which would be finished some years earlier. Ted Wilson remembers that, in view of this strong opposition from the physicists, John used the lunch break to re-examine the three-dimensional cardboard model he had made of the molasse surface and he decided that he could increase the diameter to 2.2 km. This involved taking a risk in extending the ring to include an area where there were some doubts about the quality of the rock. After a hasty consultation with Ted, John put forward this new proposal at the afternoon session, stating that it would be possible to go up to 400 GeV later with this extended design. As a result of

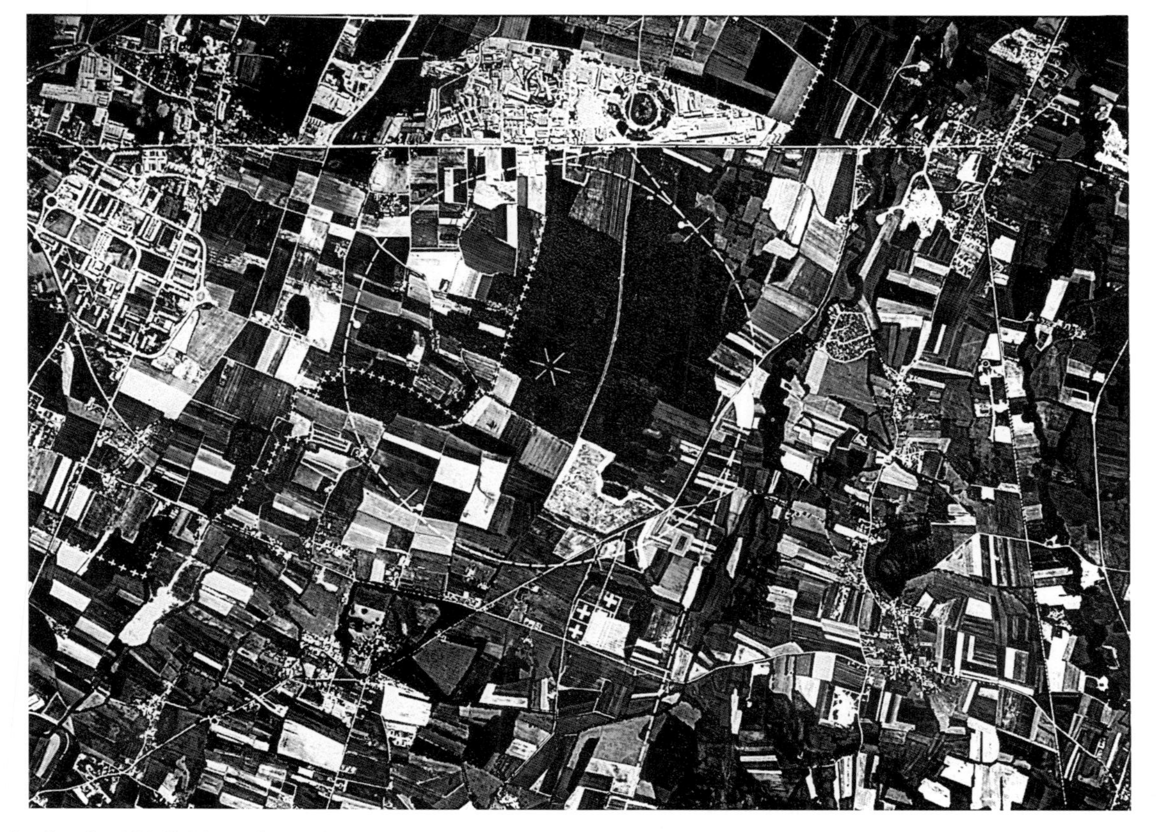

The proposed site for the 300 GeV machine alongside CERN. The existing PS and ISR rings can be seen below the main road from St. Genis (left) to Meyrin (right). The Swiss/French border is shown as a series of little crosses.

this, the meeting gave a grudging acceptance to the new proposal, should no other route to attaining the original one prove feasible. About the same time, the 300 GeV Machine Committee was recalled, told of the changes in the project and set to work on the new design.

John was a little more cautious in the paper presented to the June session of the Council, in which he stated that, until detailed borings had been carried out, it would not be possible to say whether a machine diameter greater than 1.8 km could be accommodated on the proposed site. Fortunately, the results of these borings showed that the 2.2 km ring was possible, and the Machine Committee was asked to take part in an intensive study during the months of September and October, to produce a comprehensive report on the new design by the end of November. After frantic effort by all the participants, a design report [6] was issued on the 2nd of December 1970, just in time for the next Council meeting.

In the introduction, John wrote:

> This report describes a design for a 300 GeV proton synchotron and its laboratory which, because it is located next to CERN-Meyrin, uses major capital facilities of the present laboratory as an integral part of the 300 GeV Programme, notably, the CERN 25 GeV proton synchotron as the injector for the 300 GeV machine and the West Experimental of CERN-Meyrin as its initial experimental area. Not only does this arrangement reduce the cost of the 300 GeV Programme over the 8 years of its duration but it also allows research to start with the new facilities about three years before the end of the Programme.
>
> The studies which are described in this report were carried out by the "300 GeV Machine Committee" consisting of representatives of all the European national accelerator laboratories together with representatives of CERN-Meyrin. Only three members of this Machine Committee, all CERN staff, were engaged on the study full time; the rest had other important commitments at CERN and elsewhere. The part time members, however, engaged their own staff in the design studies so that ultimately the total number of European accelerator physicists and nuclear particle physicists associated with the study numbered over 100. This unusual way of carrying out a design study had the great advantage of involving in the work a great number of very experienced European scientists who bought to the studies many different view points.

John's love of the countryside came out later in the report, where he wrote:

> A felicitous advantage of constructing the machine in a tunnel bored in the underlying rock many tens of metres below the surface is that very little disturbance is caused to the appearance of the site. All that will be visible on the surface will be the buildings in the Laboratory Area and the Plant Rooms above the long straight sectors and these will hardly be noticed in such a large area. Thus the character of the countryside of the site will remain substantially unchanged and the present use of the land for farming and forestry can continue in the future over the vast majority of the site.

The reaction of Britain to the new proposal was crucial. For the project to go ahead, it was necessary for the previous negative attitude to be changed. John had the support of the Science Research Council (SRC), represented by the recently knighted Brian Flowers (later Lord Flowers) and Gerry Pickavance, who was now Director for Nuclear Physics and had recently been elected Chairman of ECFA. John could rely on both of them to exert pressure in high places and to use their influence to rally the doubters. A new Conservative government had just been elected, and the

Margaret Thatcher at CERN. Left to right – Allaby, Mrs. Thatcher, John, Pickavance, Jentschke, Flowers, Shaw.

Prime Minister, Ted Heath, was making overtures to join the European Community, so a change of attitude to such projects could be made without loss of face.

Margaret Thatcher, Secretary of State for Education and Science in the new government, came to visit CERN in September. All the stops were pulled out for the visit and in the plane on the way over Brian Flowers gave her a tutorial on nuclear physics and the way experiments were carried out. This she took in to such effect that, when shown round an experiment using a bubble chamber and being given an explanation how it worked, she cut short the speaker, to his consternation, saying "I know how it works and I'll tell you" and proceeded to do so! John accompanied her and explained all about the new proposal. He certainly made a good impression, because on her return she wrote thank-you letters, not only to Bernard Gregory as Director-General of the laboratory, but also to John, in which she wrote:

> I wanted you to know how much I enjoyed my visit to CERN and to thank you for helping to make it an outstanding occasion. I found the work both fascinating and impressive, not least the new proposals that you outlined to me. My discussions helped me to catch something both of the intellectual excitement of the work being done there and the remarkable spirit of collaboration. I am sure that to a great extent these happy consequences flow from your work and contribution and I would like to offer you my own personal congratulations.

During her visit, she gave the impression that pure science was back on the table for discussion in Britain.

In the case of the two countries that would become Host States if the proposal was accepted the reaction was mixed. The French were not too pleased, thinking that it would involve considerable expense and administrative effort, while the majority of the benefits would most likely go to Geneva. On the other hand, the authorities in Geneva were relieved, as the running-down of the present CERN, if a second laboratory had been built elsewhere, would have meat a loss of employment and money. After some discussions, both countries agreed to make the required land available and France would supply the electricity and Switzerland the cooling water needed.

In Britain, the SRC had been reviewing the national programme and, after getting agreement amongst most of the nuclear physics community that the 300 GeV programme was of paramount importance, it produced a plan involving the eventual shutdown of NINA and Nimrod, in exchange for Britain joining this programme. This proposal was put to the Cabinet by Margaret Thatcher and, on the 4th December, she was able to announce in the House of Commons that:

> We have decided that the U.K. should participate with the other European countries which are members of the European Organization for Nuclear Research, CERN, in building a 300 GeV accelerator near the existing CERN site at Geneva. A careful appraisal of priorities within the civil science budget has made it possible to meet the cost of the project without additional public expenditure.

She had previously telephoned John as soon as she came out of the Cabinet Meeting, at which the participation had been agreed, to tell him the good news.

The agreement to participate was conditional on a minimum number of other countries also joining in, but the scene now seemed set for a decision to be taken at the December Council meeting, now that France, Germany and Italy had been joined by Britain. Belgium, Switzerland and Australia who were prepared to agree at that meeting, but Norway, Sweden and Holland were not. It was not until the adjourned session of Council, held on the 19th of February 1971, that ten Member States signed an agreement for an eight-year programme for the construction of a proton synchrotron and associated experimental facilities, at a cost of 1150 million Swiss Francs.

One other point that was to have far reaching implications was settled at this extended Council meeting; that of John's position. The Council had appointed John as Director-General designate of the new laboratory, assuming that it was going to be somewhere else in Europe. Now it was to be adjacent to the Geneva laboratory, using some of its facilities. The logical thing would be to combine the two laboratories under one Director-General. However, Willi Jentschke, from the German laboratory DESY, had recently been elected Director-General of CERN for a period of five years. The Council decided that it could hardly demote John, just at the moment it was giving him the tremendous responsibility of building the new machine, by making him a Division Leader under Jentschke, so they decided that there would still be two laboratories, CERN Lab I at Meyrin and CERN Lab II at Prévéssin, about three kilometres away, each with its own Director-General, and that the two laboratories would merge into one, with a single Director-General, after completion of the new machine.

Up till this point, the whole project had been known as the 300 GeV Programme, and the machine itself had not been given a name, but now it began to be called the SPS, which Ted Shaw, then head of the Public Information Office, suggested could stand for Super Proton Synchrotron, or Superconducting Proton Synchrotron, if that option were to be taken later.

References.

1 *Report of the Working Party on the European high energy accelerator programme.* CERN, Geneva, June 1963, (FA/WP/23/Rev. 3)

2 Johnsen, K. (Ed), The Design Study of a 300 GeV Proton Synchrotron, CERN, Geneva (CERN/563), 1964.

3 *Proc. 9th. Int. Conf. on High Energy Accelerators,* CERN, Geneva, 1971.

4 Adams J. B. and Wilson, E. N. J., *Design studies for a large proton synchrotron and its laboratory.* CERN, Geneva, (CERN 70-06), eb. 1970.

5 300 GeV Machine Committee: *A design of the European 300 GeV research facilities.* CERN, Geneva, (MC60), 1970.

Chapter 11: THE SUPER PROTON SYNCHROTRON

When the Council gave its assent in February 1971 to the revised 300 GeV project, the SPS machine and the new experimental facilities, it approved the duration of the project and its financial envelope. The participating States bound themselves to stay with the project for the full eight years at an agreed overall cost and yearly profile of expenditure. This was despite the fact that the project was not fully worked out. There were several options open for the SPS, and the experimental facilities were only defined in outline. The first job to be done was to refine the design of the SPS, which so far had been largely the result of the study groups composed of people who came from many laboratories. The definition of the experimental facilities could come later in the programme.

John started by recruiting the Group Leaders who would be responsible for the design and construction of the various parts of the machine. First of all, he looked for suitable people within the existing CERN organization, especially from the ISR Division, as this machine had just been completed. Some agreed to transfer, but others were persuaded by Kjell Johnsen to stay with the ISR and its future development. In the end, three Group Leaders came from the ISR. The first was Baas de Raad, a Dutchman who came to CERN in 1954, "whose big and ungainly appearance disguises a stickler for detail, a man never satisfied until he is doubly sure that what he has proposed will fulfil the specification"[1], to deal with the problems of getting the beams of protons into and out of the machine. Next was Simon Van der Meer, another Dutchman, but the opposite of de Raad, being reserved and not offering his opinion until it was sought. He had been responsible for the incredible accuracy of the ISR power supplies and agreed to do the same for the SPS. As we will see, he was later to be responsible for one of the most important inventions made at CERN. The third was Hans Horisberger, who brought all the attributes of a serious Swiss engineer to the problems of the Mechanical Design, a task he had performed previously on the PS as well as on the ISR.

Four Group leaders came from the rest of CERN. Klaus Göbel, a German who has been described as having "an almost old-world politeness for one looking so young", had been helping with the radiation calculations and now took over the responsibility for ensuring that it was kept within bounds. Clement Zettler, who had come from Munich to work in the ISR, where his innovative ideas were not always appreciated, was given the opportunity to show his mettle in designing the radio-frequency system for the SPS. Jean Gervaise, "a tall, articulate Frenchman with a fondness for pontifical statements", whose sensitive surveying methods had been able to demonstrate the tilting of Europe when the tide came in on the West coast, took on the responsibility for the survey work of both Lab I and Lab II. The design of the new experimental areas was not urgent, so it was agreed that Georgio Brianti who, after dealing with the PS magnet production, had shown himself to be expert in administration and personal relationships as well as in physics, would take on this work, but only part-time at first.

The remaining Group Leaders had to be recruited from outside CERN and John's first step was to discuss the problems with those that had taken part in the design studies. The first of these was Hans-Otto Wüster, a heavyweight in every respect, who could change in an instant from being his normal jovial self into a simulated table-bashing rage when encountering obstructions in a meeting. It was as a member of the Directorate of the German DESY Laboratory that he had swayed the stormy ECFA meeting and proposed the form of the resolution that was accepted. John judged him to have the right qualities to become his deputy and to look at the overall problems, and he accepted.

The magnet system was the largest component of the machine and John entrusted this to Roy Billinge, an ambitious extrovert who had worked on the earlier design studies at CERN but, impatient with delays in approval of the project, had gone to what was now called the Fermi National Accelerator Laboratory (FNAL) where he had completed his work of building the synchrotron injector. He caused some consternation by proposing to import some of the design principles used at the FNAL machine, where, due to Bob Wilson's insistence that all safety limits should be cut to the bone, large numbers of the magnets were failing, the insulation going down in the damp atmosphere of the tunnel. Robert Lévy-Mandel, described as "a compact hearty dynamo, bouncing with energy and good cheer and packing a back-slap liable to put the unwary on the floor", agreed to come from the Saclay laboratory, to deal with the general Site Installation work. Finally, John asked me if I would take on the design of the control system for this machine. Since my pet project at Daresbury, an expansion of the electron synchrotron to 15 GeV, had been turned down in favour of a concentration of the British high-energy physics resources at the Rutherford Laboratory, I agreed to take this on.

John showed great foresight in recruiting André Klein, the *sous-préfect* of Gex, the surrounding area in France, as head of Administration Services, as he was able to smooth out many problems with the local residents. Although it was planned to bore the tunnel deep under the ground, a French landowner has rights over the land under his ground, right to the centre of the earth. It was also necessary to prevent large scale building over the area adjacent to the ring, since the molasse was elastic, and a big change in the surface loading might cause deflection of the tunnel, where the magnets had to be aligned to a tenth of a millimetre. There were about 600 landowners involved, but all the necessary formalities were achieved without any major legal difficulties.

These Group Leaders, together with a small Directorate section and Ted Wilson, became the Management Board for the project, and John started his usual Monday morning meetings as soon as they could be gathered together.

The most urgent requirement was to design the laboratory and office buildings. The initial staff were housed in barracks on the Meyrin site, but these would soon be inadequate. The first experiments, with a beam energy of 200 GeV, were to be performed in the existing west experimental area of the PS, but a new area was needed for the higher energy beams planned later. The layout of the machine meant that this area had to be the other side of the ring from the Meyrin site. John decided to put the laboratory and office accommodation near this North Experimental Area, which was not far from the small village of Prévessin, so that Lab II became known as the Prévessin site. This was done in the expectation that, once the machine was complete, the major experimental programme of CERN would move to this area, which would overcome some of the overcrowding of the Meyrin site. However this was not to be. Despite agreement with the French and Swiss Authorities for the construction of a tunnel between the Meyrin site and the Pays de Gex, manned by CERN personnel, mainly for the transfer of goods and equipment to avoid customs difficulties, the 2.5 kilometres between the two sites remains today a psychological barrier and the experimenters using the North Area consider the Meyrin site their natural home.

In retrospect, due to the subsequent difficulties in integrating the two laboratories, some people have asked why it was so essential that there should have been a Lab II. For example, Kjell Johnsen asked why did John put such weight on having the title Director-General when he could have built the SPS within the framework of CERN without losing any of his autonomy of action as he, Kjell, had for the construction of the ISR. One reason was that John did not like the way CERN was organised at that time. In particular, while he was in favour of the safety factors built into what was an entirely new type of machine, he thought that the ISR Division had been built up into a rather lavish empire within CERN, with no expense spared and duplication of services that could have been provided from other Divisions. He hoped to create Lab II on more austere lines, using Lab I services where possible, but it inevitably led to some of the duplication he was so critical about. How much of the desire to have a separate laboratory was due to rational assessment and how much to the feeling that, having reached the levels he had, anything less than a Director-Generalship was beneath him, one can only speculate. One can also speculate whether the integration problems would have been any less, if the offices and laboratory had been built on Swiss territory, across the road from the Meyrin site, where some of the SPS service buildings were erected and where sufficient space could have been provided.

John now had the opportunity to express his ideas on architecture starting from scratch on a green field site. He started to plan the buildings with his usual enthusiasm and feeling for the environment. A large hall was needed for the modelling work initially and later for assembly and testing of components, especially the magnets. This was accompanied by three cross-shaped laboratory and office buildings, three stories high, so that they did not rise above the local tree level. The idea was to put communal services in the centre of each cross, so that people of different groups would meet more easily. This was true of each individual cross, but it was not until a gangway was put later between two of the crosses at the third floor level that communication between them was encouraged in bad weather conditions. The third cross building, separated somewhat further from the other two, was for the people preparing the experimental facilities, but it also housed the local canteen, so, in bad weather, we often used to go there by way of the underground ducts which carried the pipes and cables.

Every group had to justify the allocation of space that it requested in the new buildings and if John thought that any group's case was not sufficiently good, its allocation was cut until more evidence could be produced. This attitude kept the laboratory size to a minimum and emphasised the need to economise where possible. Of course this meant that some buildings had to be extended later, where the economy had been too great, but the overall effect was beneficial. John was also insistent that no trees should be cut down unnecessarily, so he made it a rule that his personal authority was needed for each felling. The North Experimental beamline was even diverted slightly from the original design because it then reduced the number of trees that would have to be cut down. With this firm insistence from the top, the planners and contractors soon got the idea that care for the environment was not just a slogan.

John would have been horrified if he had lived to see how the Electricité de France have recently cut a huge swathe through the trees and erected a massive line of ugly pylons right across the site to link the French and Swiss electrical power systems.

The other high priority item was to produce a definitive design to be presented to the Council for approval at its December meeting in 1972. The previous design had been worked out by a committee: now the people who were going to be responsible for the various components had to look into their detailed design. Some copies of the new design report were produced in time for the Council meeting and it was published in January 1973.[1]

The two laboratories officially had equal status with regard to the Council and its Committees. They had totally independent budgets but, in order to save unnecessary duplication, some services were supplied from Lab I to Lab II for an agreed annual fee. These included purchasing and financial services, recruitment and personnel administration, library and computing services, and so on. In some cases, personnel from Lab I were lent to Lab II to build up the services there, with the addition of freshly recruited personnel. As stated above, the Survey Group was common to both laboratories, and almost the whole of the effort of the Site & Buildings Division was diverted to Lab II, as there was then little work for them to do on the Meyrin site.

The fact that the two laboratories were officially equal resulted in some upsets. John caused one by demanding that his Group Leaders should have equal privileges and status to that of the Division Leaders in Lab I; not only demanding it, but getting it! Most Divisions in Lab I were responsible for more staff than the whole of John's team at that time, so the resentment was understandable. John also organised his laboratory in the way he thought best, making his own rules and regulations, where they had not been laid down by Council, and even advocated that some of them should be adopted by Lab I. One of these was the staff assessment scheme, whereby supervisors had to make a detailed report each year on the ability and performance of each of their staff, which was not then done in Lab I. In the relation between the direction of the two laboratories there was a strong contrast. On one side was John, with a clear view of his aims and the will and energy to push them through: on the other side was Willi Jentschke, an eminent scientist and a charming man, but not prepared to fight against the combined strength of his Directorate, which was largely composed of members of what Amaldi called the "Praetorian Guard", who wanted to maintain the status quo. Seen from their point of view, as expressed by Mervyn Hine, "John was really establishing his existence over there and if anything could be done differently from the way it was done previously, then John would want to do it differently, just to establish his own position."

The story of the building of the SPS has been well told in the book "Europe's Giant Accelerator"[2], so only the events that show some light on John's character will be dealt with here.

As soon as the first two of the cross buildings was finished, the majority of the team recruited so far moved over to the Prévessin site, and work began in earnest. In his usual way, John wanted to know all that was going on in every part of the project. Lévy-Mandel emphasised this point, saying:

> One aspect of John was that you could hardly ever tell him anything new; he knew it always before you did. When you came into the laboratory in the morning and you wanted to inform him about something, he was already informed because he had the habit of running around in the field before going into his office. He discovered things before you could, or would. He had an extraordinary eye to see things; nothing escaped him.

The SPS was the first big accelerator to be built deep underground and this required a total of 10 kilometres of tunnelling, mostly for the machine ring itself but also for the injection line from the PS and for the extracted beam lines to the experimental areas. It was decided that, both to avoid disturbing the residents above the tunnel line and to avoid producing fissures in the molasse that might allow water to penetrate, blasting would not be allowed and that the tunnel would be bored by a "full face" boring machine; that is one with a boring head the full size of the tunnel diameter, which is slowly rotated and pushed forward to cut into the rock. Lévy-Mandel went on:

The machine that bored the SPS tunnel.

> The other point that always impressed me was that when we started the civil engineering, he came to me on a Monday morning, after one of his weekends when he went walking or skiing 20 or 30 kilometres alone, with a complete planning drawn up. He was able to understand that kind of work without having had the opportunity before to see how a tunnelling machine worked, but he looked into it and came back with a very reasonable planning.

The tunnel was completed only about six months after the original completion date, which was remarkable for any tunnelling project, but had only been achieved by constant pressure on the consortium of three European firms that had undertaken the contract. As soon as one sector was completed, in November 1974, the installation of the thousands of components started, beginning with the dipole magnets that provide the magnetic field to guide the protons round the ring. Separate contracts had been let for the supply of the iron cores and the copper coils and the magnets were assembled, with the vacuum chambers inside, in the big hall on the Prévessin site and thoroughly tested before being stacked to await their turn to be installed.

By this time, an important decision had been taken. As reported earlier, it had been agreed to wait until the manufacture of the first half of the dipole components, needed to go to 200 GeV, was nearing its end before deciding to order the second half or to wait for the promised superconducting ones, which would allow the machine to go to higher energies later. The progress on the superconducting magnets had not been as fast as some people had hoped and it was decided to order the second set of dipole components. About this decision, Godfrey Stafford, who was then Director of the Rutherford Laboratory wrote:

> My opinion now is that Adams was correct in not succumbing to the high-energy physicist's pressure and the possibility of building the world's first superconducting accelerator. Adams always adopted a cautious approach to engineering problems and in some quarters was criticised for this. Looking now at how the complex of accelerators has developed at CERN, there is no doubt that its success rests heavily on the reliability of the machines for which Adams was responsible. On the other hand, although cautious, Adams was prepared to incorporate new ideas.[3]

The magnets mounted in the SPS tunnel. The upturned ends of the coils that had the insulation problems can be seen at the lower left.

This preparation to accept new ideas was shortly to be tested. In what John later described as "black January", misfortune struck. A technician, carrying out routine tests in the tunnel, reported that two magnets were showing insulation faults that could result in breakdown when they were powered. The magnet coils had been tested to higher than their working voltage, immersed in water, after their manufacture and so this seemed incredible. The coils were insulated with glass fibre impregnated with epoxy resin, leaving out the extra layers of mica which till then had been included in the insulation of all CERN magnet coils. People at once began to accuse Billinge of importing transatlantic practices from FNAL, where the magnets that had failed were also built without mica. Billinge had insisted that the impregnated glass fibre was not only adequate on its own, but as it was transparent, defects in manufacture could not be covered up. John had agreed to this, but was now wondering if he had made a mistake. He took over the investigation to try to find the cause of the failure, but meanwhile more and more magnets were found to have insulation faults. The offending magnets were stripped down and minutely inspected. Nothing obvious was found and a clue to the cause was only given when Billinge noticed a discolouration on one of the leads coming out of a coil, rubbed his finger on it and put it to his mouth to find a bitter taste. The deposit was analysed and shown to be phosphoric acid, which had been used to clean the copper before brazing on the connectors. A programme was rapidly put in hand to investigate whether phosphoric acid could affect the fibre-glass insulation. Rutherford Laboratory chemists found that some of the fibres were hollow, and the acid could then form a conducting path inside them which could go right through the insulation layer. Not all the coils so far manufactured had been cleaned with this acid and so, after sorting out at the factory which coils had been so treated, a programme was set up to strip down all the magnets that had already been assembled with doubtful coils and replace them. This came to a total of 280 magnets, 100 of which had already been installed in the tunnel. In remaking the magnets, a layer of Kapton, an impervious plastic, was incorporated, "just to make sure". As a result of intense activity by all concerned, the time lost by this disaster was recovered and it did not delay the completion date. John himself made the symbolic brazing of the last connection in December 1975.

Many people have remarked on John's behaviour during this incident. He was very calm. There were no recriminations, no shouting, no loss of temper. He just got down to finding out the cause of the trouble and to decide what to do about it. This was one of his characteristics; the bigger the trouble, the calmer he became. This does not mean that he couldn't lose his temper; Lévy-Mandel recalls that once, when some modifications to the

John makes the ceremonial final connection in the magnet circuit.

Lab II buildings had been incorporated without his knowledge, "that was the only time I saw him angry. He went white and left the room and Boris

The Author demonstrates the control system to John.

Interested observers at one of the pre-run tests of the SPS equipment. Left to right – Wüster, John, Milman, the author, Baconnier.

(Millman) and I went to him and explained that this was only a suggestion and nothing was definite, so he came back. This was a measure of his feelings about architecture."

After the magnet problems were solved, things went smoothly and, on the 3rd of May 1976, the event described in the introduction took place, the injection of protons into the SPS and their circulation all the way round at the first attempt. This was followed by the usual celebrations, but we drank the champagne rather than sprayed it over each other, which seems increasingly to be the habit nowadays.

After this successful first test, there was a period when things did not go too well. The first attempts at acceleration showed that the beam was

Setting up for the first test.

The beam goes all the way round!

captured by the radiofrequency system but soon lost. Systematic work showed that the beam was running into trouble with resonances, which required reprogramming of the power supplies. This is done and the beam accelerated to 80 GeV, the limit of the power supplies then connected. With all eyes on the calendar, because the Council was due to meet on the 17th of June, Van der Meer and his team were working night and day to get more power supplies connected in. By the 4th of June, 200 GeV was reached with half the power supplies, then other troubles appeared, but they were sorted out and on the morning of the 17th all was prepared for a test with the full set. The control room was even more crowded than it had been for the first tests six weeks earlier. The power supplies were programmed for 300 GeV and when they had settled down the call was made for the PS to inject a beam. The beam was captured and the crowd was almost silent, watching the screens for the four seconds it took to reach peak field. The current monitors showed that some of the beam remained right up to the top. 300 GeV had been reached! Again shouts and congratulations all round.

I left the noisy throng and drove over to the Council Chamber on the Meyrin site, where the Committee of Council were in closed session, preparing for the afternoon meeting of the Council, and passed a note to John, just before the Committee broke up for lunch, so he was able to announce this achievement. After lunch, the Council assembled and John was asked to present his report on the accelerators at CERN. He got up and very calmly went through the events of the past few years and the building of the SPS, ending up with the bald statement that it had reached its design energy of 300 GeV that morning. He reminded Council that this was the highest energy that had been authorised and requesting permission to go higher. This was because the Dutch delegate, fearing an escalation of costs, had persuaded the Council to put this limitation on the energy when the project, which had not then been completely defined, was approved. The Council gave its assent and John got a message through to me. Van der Meer had already reprogrammed his power supplies and it was not long before we saw the beam go all the way out to 400 GeV. Clasping the photographic evidence, once again I drove over to the Meyrin site, only to find that this time I was just too late, as the meeting was then adjourning for tea. I gave John the picture and, when the Council re-assembled, he was able to announce the success, to congratulations all round.

This was not the end of the programme, which still had three years to run. The new North experimental area had to be built and equipped, but meanwhile the extraction equipment and beamlines for the existing West Hall on the Meyrin site could be brought into operation for experiments to start. For a number of reasons, the maximum beam energy that could be used there at that time was 200 GeV, so the SPS started its first scheduled operation for physics at that energy.

The estimated cost of the 300 GeV programme had included a reasonable contingency factor for escalation and unexpected costs, but it had been completed within the original budget, without having to call on this contingency. This meant that there was more money available for the construction of the North experimental area than had originally been envisaged, and part of this was used to extend the facilities. The rest was used later, when there was a pressure from the Member States to lower the overall CERN budget, to reduce their contributions, rather than reduce the budget.

During the construction of the SPS, John was approached to see if he might be interested in joining an industrial consortium. He wrote back to say that he was committed until 1979, but that when the machine was finished in 1976, he might be able to take on another job part-time, subject to agreement with CERN, as there would then be only the North Area to complete, and he did not expect that would occupy him full time. However that was not to be, as we will see in the next chapter.

References.

1 *The 300 GeV Programme,* CERN 1050, Geneva.

2 Goldsmith, M., and Shaw, E., *Europe's Giant Accelerator*, Taylor & Francis Ltd. London 1977

3 Stafford, G.H., John Bertram Adams, 1920-1984. *Biographical Memoirs of Fellows of the Royal Society*, **32** (1986) 3-34.

Chapter 12: DIRECTOR-GENERAL

While the SPS was coming into operation, the Council were faced with an important decision. Willi Jentschke's term of office as Director-General of Lab I was due to terminate at the end of 1976. On the other hand, John had been appointed Director-General of Lab II for a period of eight years, the duration of the project, rather than the normal five years, and so his term went on to the end of 1979. The Council faced a dilemma: it could either appoint a new Director-General for Lab I and tackle the problem of the integration of the two laboratories later, or it could face up to it now and look for an acceptable solution. There was a very strong body of opinion that the two laboratories should be joined up under a single Director-General and so an Electoral Committee was set up to advise the Council how this could be done and to propose a candidate for the post.

Obviously, the first to be considered was John: he had an international reputation and still had three years to run in his present position. The major problem was that it was thought that the head of Europe's premier high-energy research laboratory should be a nuclear physicist with appropriate and proven research experience rather than an engineer, even one of the stature of John, or an administrator. The Electoral Committee made discreet enquiries to see whether it would be acceptable to the physics community to offer John the post of Director-General of the combined laboratory, but only for the remaining three years of his present contract. The opinion seemed to be favourable, and so they approached John. If this did not take John by surprise, he kept it to himself, since in his reply to a letter from the President of the Council, Professor W. Gentner, in October 1974, he wrote:

> You will appreciate that any views which I express now can only be preliminary and subject to modification after discussions with senior staff at CERN, and with the SPS, and with Council delegates. Until last week I had the impression that you were seeking a solution to the succession of Professor Jentschke in an entirely different direction, and consequently I have not given much thought to what I would do if faced with this new and heavy responsibility....
>
> When I was asked by you and Professor Gregory, towards the end of 1968, whether I would consider returning to CERN to take over the 300 GeV Programme I was then the Research Member of the Board of the

> U.K. Atomic Energy Authority and executively responsible for the Harwell and Culham Laboratories whose total staff numbered about 6000. I was also a member of the Council for Scientific Research which advised the Minister of Education and Science and of another Council which advised the Minister of Technology. I had previously helped to set up the Ministry of Technology and was its first scientific head. In other words, I was not responsible for two large laboratories whose programmes were being seriously questioned and whose budgets were being cut down, but I was also deeply involved in government affairs concerning scientific research budgets and technological development programmes.
>
> Having worked for several years on these difficult and worrying problems and having assured, as far as possible, the future of the Harwell and Culham Laboratories by getting two excellent directors appointed for them in whom I had every confidence, I not only welcomed the opportunity which CERN offered me to return to my real professional activities but felt that I could accept it with a clear conscience.
>
> When it was suggested about a year ago that I might be interested in taking over both CERN Laboratories I felt that the work would be too similar to that which I had been doing in Britain and which I had left in order to build the world's largest accelerator. After all if I had wanted that kind of work I could have stayed in Britain where I would also have reaped the rewards which go with such jobs in one's own country. Hence I was very reluctant to consider this idea.

Here he obviously had in mind the almost automatic knighthood that goes with the sort of position he previously held. It is also clear from the above that John had by now overcome any earlier difference he may have had about his abilities, even to the extent of somewhat exaggerating his influence at the Ministry of Technology and ignoring his frustration at having had so little effective power in the Authority. He went on:

> Since then things have changed dramatically in Europe and their effects are now beginning to be felt at CERN. Without being unduly pessimistic, I think the coming years will be some of the most difficult ones for the Organisation and, as I have said, the unification of the two CERN Laboratories under one management will be a great help in facing the difficulties. Therefore, providing arrangements can be made for me to finish the SPS project, I am much more willing now to consider the solution which you are putting forward.

He would only take on this new proposal on his own terms and one of these was that he should be appointed for a full five years, the normal term of office of a Director-General. Having wasted time earlier, the Electoral Committee now pressed for a rapid response from John as to how he would re-organise CERN if he took over. In his reply to the President of the Council, he wrote that there were three specific areas of anxiety. The first was the completion of the 300 GeV project, which he expected to require half his time in 1976. The second was to build up a new management for the whole of CERN, which should be decided in 1975 in order to come into operation at the beginning of 1976, when some of the present Directors ended their terms of office. The third was the management of the research programme, where he proposed one or two senior scientists to determine the research programme under his direction. He continued:

> Regarding the duration of the appointment, I am quite sure that three years would not be long enough for me to do the job which I have in mind and which I have tried to explain in this letter, especially since during the first year I would have to share my time between the 300 GeV Programme and the management of the whole of CERN.
>
> I can well imagine the difficulties which the SPC must have gone through before coming to their present recommendation and I an deeply impressed that they have shown such confidence in someone whose main contribution to high energy physics is building accelerators. I can also easily understand the reservations that must be in the minds of many of the members of the SPC about this recommendation since I have exactly the same reservation myself. What I find difficult to understand is that having come to this rather courageous conclusion the SPC is unwilling to allow me the time necessary to do the job, which is no longer than the normal appointment period for a Director-General. The argument that I would be Director-General for ten years if I were given another five years is not tenable. The present Director-General of the vast majority of CERN is Professor Jentschke, not me, and he will remain Director-General until the end of 1975. My responsibilities during the term of office of Professor Jentschke were carefully laid down by the Council and they only concern the 300 GeV Programme. Professor Jentschke is responsible for all the rest.
>
> Rather than attempt to do this very difficult job in 3 years and not succeed, I think it would be better for CERN and for all of us that I complete the job the Council asked me to do in 1971 which can employ all my time and talents up to 1979.

After giving some of this ideas on how he would run CERN if he was appointed Director-General, he ended:

> Finally I would like to come to a matter which seems to have taken some members of the Committee of Council by surprise, namely, my negative reaction to the idea of a 3-year appointment. There are several reasons for my reaction, some more important than others, and I will mention them in no particular order. Firstly, the normal term of office of a Director-General is 5 years and if I am appointed for 3 years unkindly people will interpret this to my disadvantage, which will undermine my effectiveness. I think I am right in regarding this as a new appointment since my present post will disappear when Laboratory II is wound up and I should have to be appointed to a new post. Secondly, one can hardly expect the staff of CERN, particularly the senior staff, to pay much attention to a Director-General who is in office for such a short time. They will regard him simply as an interim solution awaiting the day when a real Director-General can be appointed. Thirdly, I shall want to continue my personal involvement with the SPS until the end of 1976 when we hope to get the machine working. This was the job I undertook to complete for CERN and my justification for leaving my responsibilities in Britain. Fourthly, the last year of office of a Director-General is never very effective since everyone is waiting for the new man and putting off action until he arrives. Lastly, and this is the most important point, I do not believe I can complete the job which I consider is necessary for CERN in 3 years.
>
> I give these reasons frankly, realising that this may upset your plans. I am sorry if this is the case, but I believe the next Director-General of CERN is going to have a very difficult job to do and should be given every chance and every encouragement to do it successfully. If the Council offers it to me and I accept I want to do it well and to leave CERN in a strong position to face the 1980s.

In fact, in November the SPC did reluctantly agree to the five years, if that was the only way out, but the members of the Electoral Committee were not convinced. If John's term of office had not had three years still to run, the Committee's choice for the next Director-General of the combined CERN would have been Leon van Hove, then head of the Theory Division, despite this being in conflict with the general rule that it was preferable to appoint someone from outside CERN. In fact the SPC recommended that if a suitable arrangement could not be reached with John, Leon should be appointed as Director-General of Lab I and the two Laboratories should be united under him three years later. There was also some worry about John's proposal to spend half his time in 1976 on the SPS and the SPC wanted to know whom he would delegate the technical responsibility for its completion.

Another idea came up and that was to merge the two laboratories, but to appoint two people as joint Directors-General, John for the management of the laboratory and Leon for the direction of the research programme. The lawyers looked into whether such an arrangement would be allowable under the CERN Constitution and decided that wording could be found for their appointment, as two persons sharing a single job, that did not conflict with it. This proposal was put to John and Leon and they were asked if they thought they could work out a satisfactory arrangement.

They had a number of conversations to discuss the way they might operate together, mainly in January 1975, and found no insurmountable difficulties. John wrote a proposal as to how they thought they could operate as joint Director-Generals and sent it to the President of the Council in February. However, John must still have had doubts about this proposal because the accompanying letter included the following paragraphs:

> Before these documents are passed to the Members of Council, I feel I must make clear my personal views on the solution which is contained in the two resolutions and the procedures by which the Council may reach its decision.
>
> In the first place it must be clear to everyone concerned that this solution is not the one I have proposed. My proposal is contained in a letter which I sent to the Council delegates at the end of last year on the request of the President of Council. I still hold that my proposal is far better than the present one for ensuring the good management of CERN. I accept that the Council could not agree to my proposal and hence could not reach a decision.
>
> In the second place it must be clear that it was the Committee of Council who proposed the present scheme of two Director-Generals and a single unified Laboratory. Van Hove and I were asked to consider if we could manage CERN on this basis and our answer was discussed at the last meeting of the Committee of Council where it was unanimously approved. Nevertheless, the responsibility for this scheme rests with the Committee of Council and of course with you as President of the Council.

This letter was received with some concern by the President and, John being away, George Hampton, then head of Administration, had to cool him down, saying that this was a personal letter from John to the President to make him aware of his views and it was not an official communication.

A *modus vivendi* was worked out and the proposal to have a double headed organisation from beginning of 1976 was accepted at a special meeting of the CERN Council held on the 21st. of March 1975, whereby John would become Executive Director-General and Leon would become Research Director-General. Over the next month or so they worked out a new organisation for CERN and the division of responsibilities.

The Meyrin Laboratory still had roughly the same organisation as when John had left it in 1961, although some changes had been made by Weisskopf and Gregory, the main one being that many of the Divisions had been grouped together into a number of Departments, each with an executive Director who was only responsible to the Director-General. John did not like this, as some Departments tended to operate as autonomous units, with many of their own services, causing wasteful duplication. Also many of the projects being undertaken required the services of more than one Department, which created difficulties in the apportionment of resources

Willi Jentschke is happy to pass over the Directorship of CERN to van Hove and John, in the presence of Paul Levaux, President of the Council.

and responsibilities. He won over Leon to his ideas of matrix management and got him to agree to a plan that the Divisions should be kept much as they were, with the addition of Lab II as the SPS Division, but with the Division leaders reporting directly to the Director-Generals. The Departments would be replaced by a Directorate of six non-executive Directors, "to establish the general policies of the Laboratory and its programmes and to ensure that the resources of the Laboratory are efficiently and economically deployed on its programmes." The horizontal part of the matrix would be provided by the appointment of a Project Leader for each major project, who would be allocated a budget and resources in the Divisions concerned for its completion. In a note to explain this scheme to the Council, he wrote:

> In order to allocate the resources of CERN in the best possible way in the future, programmes of whatever type which individually use more than some minimum amount of the resources of CERN should be treated in the same way. Each programme should
>
> – be clearly defined in terms of its objective, content, duration and the financial and manpower resources to carry it out,
>
> – be approved by the Research Board as being necessary for carrying out the research programmes of the Organization in the short or long term.
>
> – be entered in the Programme Budget System so that it can be identified and the resources it actually uses accounted for.
>
> – have a nominated leader who is responsible for its execution and who controls all the resources allocated to the programme for its duration.

John also wrote a note, which is a beautiful example of his clear and logical thinking, explaining the need for this type of organisation in more detail, which ends up:

> Lastly, the importance given in this note to the horizontal structure of the programmes and the resources deployed on them is not some kind of management gimmick, but is considered essential by both Director-Generals in order to justify the budgets and manpower resources of CERN, now and in the future, by relating them clearly to well-defined programmes whose importance to the research carried out at CERN can be explained and defended.

This change of structure was resented by some of the senior staff, including some of the so-called Praetorian Guard, who lost some of their power and influence. Their resentment showed up later, when at the end of 1979 the Council accepted the recommendation of a review board that the salaries of the senior staff should be effectively lowered by reducing, and in some cases suppressing, the annual cost-of-living increase. This result in a very stormy meeting of the senior staff, in which John was accused of not sticking up strongly enough for the personnel. He was visible shaken by the verbal abuse he was subjected to from staff who had received a very comfortable living from CERN for many years, and were considered by the outside physicists to be over-paid, although CERN salaries were actually somewhat lower than those in other International organisations in Geneva.

John must have felt depressed about this and, to boost his spirit, he wrote down Machiavelli's warning in his notebook:

> There is nothing more difficult to carry out, nor more dangerous to handle, than to initiate a new order of things. For the reformer has enemies in all those who profit from the old order and only lukewarm defenders in those who would profit from the new.

As mentioned before, the previous Director-General, Willi Jentschke, was a delightful man and a good physicist, but did not have the strong character needed to make a successful Director-General. He had left administrative matters, such as personnel, finance and relations with the Member States, to the Director of Administration, George Hampton, who had made himself a little empire. He admitted later that he had made decisions and done things that he could not have got away with under any other Director-General. Previously, John had been critical of this situation and when he and van Hove took over John decided that he did not need a Director of Administration. Hampton did not want to take a post of less responsibility, "just being left with the tedious runt of the job" as he put it, so arrangements were made for his departure from CERN. He told me that later on John admitted to him that he had underestimated the work involved on the personnel side and had later appointed a Director of Personnel to take some of the load off him.

At the time of the staff troubles, a visitor to the DESY Laboratory on meeting Jentschke asked why he looked so happy. He replied, "When I was at CERN they thought I was the worst DG they ever had ... and now they don't!"

All through his period as Executive Director-General John had to struggle to minimise the cuts in the CERN budget that the Member States wanted to inflict. In 1975 it had already been decided that the budgets, instead of increasing steadily, as they had in the past, should decrease by 3% per annum for the next three years and the Council demanded a cut in staff numbers. In the following years, some countries were in financial difficulties and wanted reductions greater than this annual 3%. John took great trouble in preparing his presentations to the Finance Committee and Council, explaining the consequences of applying cuts to the various parts of the programme. His problems were increased by the pressure from the physics side for increased expenditure on the experiments and less on the accelerators, and by outside physicists who claimed that CERN did things on an extravagant scale, but then complained when CERN did not provide them with every service they desired.

With all these difficulties, one can understand why, when I asked Leon, shortly before his untimely death, about his impressions of their period as joint Director-Generals, he said:

> The most striking ones, if I summarise now the impressions and conclusions that came to my mind, often after the fact, was that I was the happy man in the combination and John was not a happy man, but he

A gathering of CERN Director-Generals on the occasion of the 25th birthday celebrations. Left to right – John, Jentschke, Bloch, Weisskopf, van Hove.

never complained. He went to the extreme of correctly respecting the boundaries of decision making and influence. He was a man of very great self-control, of very honest and straightforward handling and I was a man of emotional, individualistic and impulsive type. I developed certain convictions and I defended them in a very extrovert way, while he was much more a man for thinking out things and communicating them by small steps. Despite all that, the thing worked.

John was happier when dealing with the technical aspects. As we have seen, he retained control of the Lab II, now the SPS Division, until the end of 1976, to see the successful start up of the physics programme in the West Area on the Meyran site and the start of the construction of the new North Experimental Area on the Prévessin site, before handing over the Division to me. He started an "Accelerator Club" which met regularly under his chairmanship to discuss improvements to the existing accelerators and ideas for new ones. As is usual at most laboratories of this kind, the accelerator experts have hardly finished one project before they are thinking of the next. All sorts of possibilities were discussed, from increasing the collision energy in the ISR, by using superconducting magnets, to colliding electrons and protons in the SPS, and design studies were published under acronyms such as SCISR, MISR and CHEEP.

The most interesting of these proposals, both from the physics and accelerator experts' point of view, was the result of an invention by Simon van der Meer, who had been responsible for the power supplies for the SPS. This was a method of reducing the size of beam of particles, know as "stochastic cooling", which had been demonstrated to work on the ISR. This was seized on by an experimental physicist, Carlo Rubbia, a massively built extrovert who then seemed to do most of his work on transatlantic planes, holding simultaneously posts at Harvard in the USA and at CERN. It was well known that beams of particles and their anti-particles could circulate in a machine in opposite direction to provide collisions with twice the energy of each beam and that anti-protons could be produced by protons hitting a heavy target, but only in small numbers and in a very diffuse beam. Rubbia proposed to use van der Meer's idea to store and "cool" a beam of anti-protons, produced by successive pulses of PS protons, until it had a high enough intensity and small enough size to inject into the SPS, to collide with a contra-rotating beam of protons. The aim was to produce collisions with enough energy to produce two new particles, know as the W and the Z, whose existence was necessary to provide the next stepping-stone in the search for the "Great Unified Theory" that would link all the nuclear forces together.

A committee was set up to look into this possibility, which Leon thought most important. As he told me:

> To me, this project, from the very moment I heard about it, which was very early in our period, was the saving element for CERN, because I was absolutely convinced that, despite all the perfectionist qualities of the SPS and of the detectors, which were very, very complete, very powerful detector systems, we were going to do the physics of the second generation and it was, of course, already rather clear scientifically, that physics frontiers, as in the past, were not going to be reached that way. So the proposal of Rubbia was tremendously welcome and then came this debate, where again John was absolutely correct – I think he did not favour that project. Once he told me that he would favour an energy doubler, a good one.

The energy doubler would involve changing the SPS magnets for superconducting ones with at least double the magnetic field, as was being done in the USA at the Fermi National Accelerator Laboratory. Van Hove's reference to "a good one" was the result of the difficulties then being experienced with the superconducting magnets in America.

Once John had been convinced of the possible outcome of the anti-proton proposal, he authorised an experiment, using some magnets from a former experiment, to confirm the calculations on cooling. He also set up a "p-pbar Steering Committee" (p and pbar being the symbols for the proton and anti-proton), to look into all aspects of the proposal, which would require modifications to both the PS and the SPS as well as the construction of a "cooling" ring, called the Anti-proton Accumulator (AA). Despite the fact that Johnsen was free and might have expected to have been offered the job of project leader, John did not appoint anyone, as he would have liked to have taken that position himself if he had not been Director-General. Instead, he gave the responsibility for the PS and SPS parts to the respective Division leaders and set up a new Group, under Billinge and van der Meer, to design the AA. All the major decisions were made by the Steering Committee, with Franco Bonaudi, then Directorate Member for the Accelerator Programme, in the chair. John, as a member of this Committee could then use his influence to steer things the way he wanted them to go. For example, there were arguments about whether the anti-protons should be injected directly from the AA into the SPS at 3 GeV or sent back to the PS and accelerated to 25 GeV before injection. The arguments were fairly evenly balanced, on one side a greater risk of losing antiprotons and on the other, greater complication in operation and increased cost. By then I had taken over from Bonaudi, and I prepared a list of all the advantages and disadvantages of the two schemes. John took it home at the weekend and came back the following Monday and said "since no-one seems able to make a decision, I have taken it on myself". Characteristically he had chosen the lower risk option.

By the beginning of 1978 a design for the AA had been worked out as well as plans for the rest of the project, the cooling experiment had been successful and everything was ready to go. Normally, a project of this size would have had to be submitted to the Member States for approval and budget allocation. However, since it was split up amongst the Divisions, and the estimated cost of the AA was not all that much more than that for a major experiment on the SPS, John and Leon managed to get it accepted by the SPC and Council as a logical development of the CERN facilities and that it could be carried out within the authorised budgets.

Once approved, the work proceeded rapidly, thanks to the by now very experienced teams in the Divisions. The first anti-protons were injected into the AA in July 1980 and the whole complex came into operation with proton-antiproton collisions at 270 GeV in the SPS in 1981. Meanwhile, Rubbia had been directing the construction of his experiment in one of the two enormous underground halls which had been excavated round the SPS tunnel. The first W particle was detected in 1982 and the first Z was identified early the following year. The successful result of this project was recognised by the award of the Nobel Prize for Physics to Rubbia and van der Meer in 1984, which must have been one of the most rapid awards of all times, since most Laureates have to wait many years for their contributions to be recognised in this manner. Unfortunately, John was not to take part in the ceremony, as he has died shortly before the award, but in the speeches it was said "The late John Adams had been responsible for the construction of the two outstanding proton accelerators which were called into action in new roles for this project".

Meanwhile, discussion had been going on about the next big project for Europe. In March 1977, a meeting of the European Committee for Future accelerators (ECFA) came to the conclusion that this should be a ring in which electrons and positrons (anti-electrons) collided with an energy of at least 100 GeV in each beam. The machine experts at CERN had been looking into the feasibility of designing a Large Electron-Positron storage ring (LEP) since early in 1976 and had found some technical difficulties with the 100 GeV machine they had been studying and the investigations were transferred to a smaller machine, called LEP-70, since the first stage energy of this design was 70 GeV. Just as superconducting magnets allowed higher magnetic fields to be produced, using superconducting radio-frequency cavities, instead of conventional copper ones, would allow greater electric fields to be produced for the same RF power. Experiments were going on at a number of laboratories, and some successful cavities had been produced, but there were still difficulties in obtaining repeatability. Assuming these could be overcome, LEP-70 should have been able to go up to 100 GeV, using superconducting cavities. A design study, known as the "Blue Book", was published in 1978.

John was reluctant to enter the more speculative area of the electron machines and would have preferred a big proton machine, using superconducting magnets. His notes written about the ECFA meeting even include the statement "ECFA's mind (is) made up and doesn't want to be confused by facts"! However, by July he had been convinced that the electron machine was what the physicists wanted at the time, but thought that it provided a very specialised set of facilities and "what of the 1990s?" He therefore wrote a note proposing that the tunnel for the electron machine should be made large enough to leave room to fit a ring of superconducting magnets to enable protons to be accelerated to above 3 TeV. He called this combination SPEC (Super Proton Electron Complex).

The design of Lep-70 was discussed at meetings of potential users and members of ECFA in late 1978 and early 1979. The LEP Study Group were urged to look again into the possibilities of a larger machine, which could reach 100 GeV per beam without the need for superconducting cavities. This was to ensure that it would be possible to produce pairs of W particles, whose mass could only be estimated at that time,

The result of these further deliberations was published in August 1979 as the "Pink Book". It proposed a LEP of 30.6 km circumstance, to be built in stages according to the amount of RF equipment provided. Up to 100 GeV per beam could be provided with copper cavities and this might be increased to a 130 GeV with superconducting ones. Earlier doubts about the technical problems with such a large machine had been resolved.

After the publication of the "Pink Book", the CERN Management began to prepare the ground for a formal proposal to the CERN Council to build such a machine. To keep the original cost down, it was proposed to build the machine in stages. All the magnets would be installed from the start, but only one third of the radio-frequency equipment, which would allow an energy of 62 GeV per beam to be attained. This was known as Stage 1/3 and was estimated to cost 1064 MSF.

The proposal was discussed in the SPC, where some members thought the cost was too high. Herwig Schopper, who was then Director of the German DESY laboratory, where a machine to collide electrons with protons was being planned, claimed that the CERN design was too expensive and that DESY could build such a machine for much less. By then, I had been appointed Directorate Member for the Accelerator Programme and John and I spent a lot of time going over the various possibilities for reducing the cost without sacrificing the good engineering that had resulted in the rapid commissioning and reliable operation of the previous

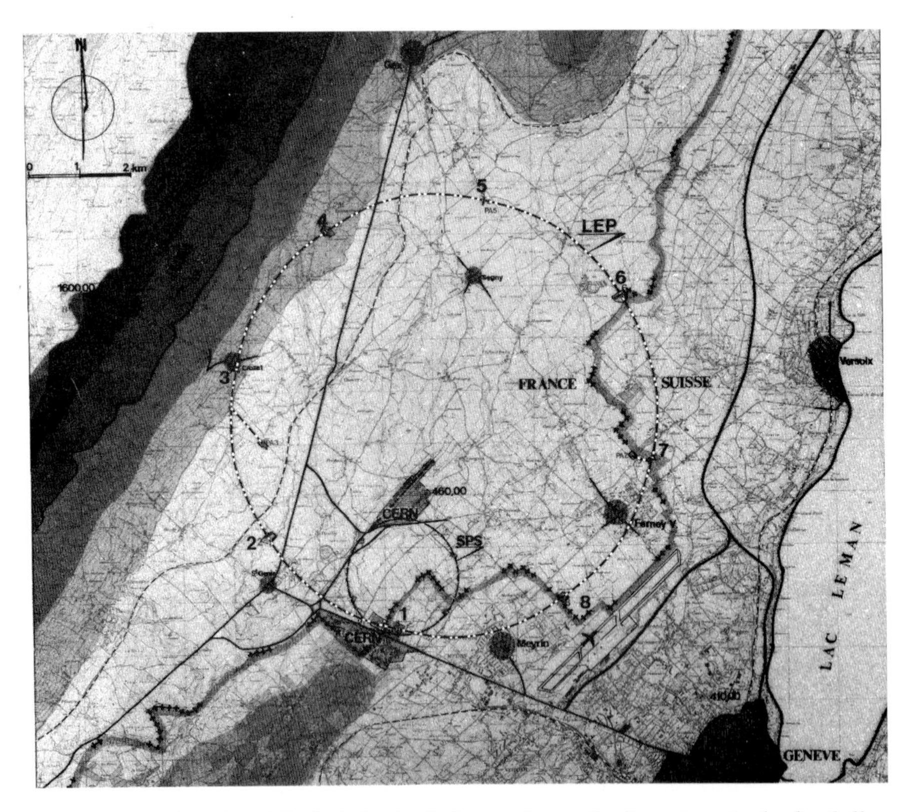

Site of the LEP project. The lightly shaded area shows the limestone in the foothills of the Jura mountains.

CERN machines. The amount of RF equipment was further reduced (Stage 1/6), the number of experimental areas to be excavated initially was reduced from eight to four and the capacity of the electricity supply and cooling systems was reduced. The possibility of using the PS/SPS complex to accelerate electrons and positrons, to replace the injector synchrotron was investigated in April 1980 and adopted by June, when the project became known as LEP Phase 1.

The LEP Study Group continued to look into detail improvements in the design and the advice of experts from other laboratories was sought "on the best technical design of LEP and its components which will enable LEP to be built and commissioned as fast as possible and at minimum cost" by the formation of the LEP Machine Advisory Committee (LMAC) in April 1980. One of the worries was that, to accommodate the required size of tunnel near CERN, part of it would come out of the good molasse rock and into the limestone at the base of the Jura Mountains. This limestone could contain cavities filled with water at high pressure. To minimise this risk, a re-alignment of LEP with respect to the SPS was proposed, together with a reduction in the circumference to 26.6 km and the tilting of the tunnel from the horizontal. This cut the length of tunnel to be excavated under the Jura from about 9 km to 3.3 km as well as reducing the maximum depth of the machine below ground level. This change also allowed the cost estimates to be reduced still further.

Meanwhile, the Council had decided to appoint Herwig Schopper to be the next Director-General of CERN from the beginning of 1981, when John and Leon came to the end of their term of office. He picked on Emilio Picasso, an ebullient Italian and a prominent physicist, who had carried out many important experiments at CERN, to be project leader designate for LEP, leaving John with no further part to play in this project. However, John and Leon had done the groundwork so well, including getting the Council to agree that LEP, if approved could be treated as an extension to the CERN basic programme, rather than calling for the extensive discussions that would have been required in the Member States for the approval of a new programme, so that when Schopper went to the Council in June 1981 with the formal request for permission to build LEP, it was received favourably. The project was approved at the following meeting in December, with the condition that it should be built without any increase in the CERN budget and that savings should be made by shutting down the ISR.

Since John had an indefinite contract with CERN with several years to run, he was given the title of senior scientist and his old office in the SPS buildings at Prévessin, but no real duties. It is somewhat ironic that John, who had progressively reduced the influence of Johnsen and Hine after his return to CERN, now had the same experience: the biter bit!

This sudden fall from the power to determine the direction of CERN came as a severe shock to John, who expressed to Robert Lévy-Mandel how difficult it was to go from one situation where you had a queue before your office to one where no-one would step in from one day to anther. However, he was not entirely neglected. Schopper had replaced all the members of the Directorate, and I was asked by Picasso to look into the control aspects of LEP. Despite our offices now being at opposite sides of the two CERN sites, John and I used to meet for lunch whenever possible.

Although John's talents were no longer fully used at CERN, he was still in demand elsewhere, as we shall see in the next chapter.

Chapter 13: *THE INTERNATIONAL SCENE AND THE LAST YEARS*

Throughout his life, John took every opportunity to foster international collaboration in science and technology. This was one of the things that attracted him to CERN and we have already seen that, as early as 1964, John was amongst those putting forward the proposition that the construction of a very big accelerator might have to be considered as an intercontinental project. This was somewhat premature, and the possibility was not brought up again until some years later, but collaboration at a less ambitious scale was encouraged. As we have seen in the early CERN days, the exchange of information on accelerators between Europe and the USA on accelerators was simple and effective, but the situation was different with the USSR at the start. It was only in 1955, at The International Conference on High Energy Physics in Geneva that any details of the Russian accelerator programme were revealed, with the first Western scientists being allowed to see the existing machines at a conference in Moscow the follow-

John and Bas Pease with Artzimovich and other Russian delegates during the conference on fusion research in the USSR.

ing year. After that, despite the cold war, collaboration between CERN and the USSR extended to carrying out joint experiments on the machines in the two areas.

This collaboration extended to the fusion field, and John led a delegation from Culham to a conference on this subject that was held in the USSR in 1968, following which an exchange of scientists between the two countries was arranged.

In 1969, when what was then to be the largest accelerator in the world was being built at Serpukhov in the Soviet Union, CERN agreed to build the equipment to extract the beam from the accelerator, in return for the use of the beam for some CERN experiments. The team that went from CERN to install this extraction system got their first experience of what it was like to work in Russia at that time. The Russians were all very friendly and helpful, but the bureaucracy was impregnable. John was to find out this for himself when, on a later visit, he was held up for eight hours at the Moscow airport because his visa did not have the stamp of one of the hotels he had stayed at, despite the fact that he was a VIP guest of the Academy of Sciences. After that, when CERN was asked to send a representative to a conference or other official meeting, John used to say to me "it's your turn to go, Michael"! Another example was when we both went to a conference at Serpukhov. He, as Director-General, was given a three bedroom dacha and I, as a mere Division Leader was given a small flat. When John said "this is nonsense, why don't you move in with me, there is plenty of room" it was pointed out that this was against the regulations, one had to sleep in the accommodation allocated, but I was allowed to dine with John, who had also been provided with a cook to minister to him!

It was not only in the USA and USSR that John went to spread the gospel of international collaboration. When the first opportunity came to visit China, in September 1977, John, Renie and I went to Peking, on the invitation of the Academy of Sciences to advise the Chinese on their plans to enter the high energy physics field by building an accelerator. The head of the Institute of High Energy Physics, Chang Wen-Yu, at that time in his seventies, had been a student at the Cavendish Laboratory, Cambridge, under Rutherford. At first, our hosts seemed somewhat reserved and kept the conversation to generalities, but the following day the historic proclamation by Teng Hsaio-ping was published, announcing the decision that no longer would China remain isolated but would be open to use the best of western technology. After this, our hosts said that they could now talk

Sign outside a Chinese factory. It reads "Welcome to our uncles and aunts of the European Organization for Nuclear Research who have come to visit our children's crèche".

more freely. One thing they wanted to know was if we thought that their industry could produce the components needed to build such a machine and we were taken round various factories. We were horrified to see, at a factory making electric alternators, workers stamping out laminations from steel sheets which they fed by hand into giant presses with no guards whatsoever. Life was cheap! At each factory there was a painted sign outside to welcome the foreign visitors and the children in the factory creche performed a little dance for us.

A site outside Peking, near the Ming tombs, was chosen for the new accelerator, which was to be built in stages, the first to be a small proton synchrotron, which later would be used as injector to the second stage which was planned to rival the SPS. At the end of our visit we were invited to an audience with Teng Hsaio-ping at the People's Palace. He had obviously been kept aware of our activities and the discussion lasted for an hour. He said:

Audience with the Chinese Premier, Teng Hsaio-ping.

> China's aim in developing science and technology is to strive to come close to the advanced world level in science and technology by the end of the twentieth century, to reach that level in some fields and to go beyond that level in certain particular fields. We must have our share in discoveries and make our contribution to mankind.
>
> We have overthrown the "gang of four", mobilised the energies of scientists and technologists and of the masses of the people; we are therefore really confident that we can achieve our aim.

By the time John paid his second visit to China, two years later, they were still planning the first stage of the accelerator and Owen Lock, who accompanied him on this trip, remembers that "we spent a whole morning looking at the site, with John asking questions about the geology and stability and so on. It was a beautiful sunny day in May and John was puffing his pipe and just interested in the challenge of building a PS out there in the middle of China". Shortly afterwards, they decided this was too ambitious a project for the resources available and decided to concentrate on a smaller electron/positron collider, which could be built alongside their existing laboratory.

This second visit to China was a stepping stone on the way to Japan, where John wanted to spread the gospel of collaboration, especially the exchange of personnel between CERN and the Japanese laboratories and the possible support of the LEP project by Japan. Although earlier thoughts

The second visit to China. Discussing the Ming tombs site.

of an International accelerator had been premature, the subject continued to be discussed at various International meetings. At one such meeting in 1976, at Tbilisi in the USSR, it was proposed that an International Committee for Future Accelerators (ICFA) should be formed, with the brief to foster collaboration between the regions and to look into the requirements for a world scale machine, the Very Big Accelerator (VBA). John was a member of ICFA from the start and, on the untimely death of the first chairman, Bernard Gregory, he was elected to replace him at the second meeting. He remained in that position for the next five years, in recognition of his pre-eminent position in fostering such cooperation. Although we are still far from seeing the possibility of a world machine, the Americans having decided that they can build such a machine on their own, ICFA was successful in defining the conditions for the use of national machines by physicists from other continents and in sponsoring meetings to discuss the new techniques that would be needed for future accelerators.

Some time before John's period as Director-General of CERN came to an end, there were people clamouring for his services. He was approached in April 1979 to see if he would be interested in the post of Director-General of the European Space Agency (ESA), as this would become vacant in April 1980. Reporting on a meeting in Paris where this proposal was put to him, John wrote in his notebook:

> My reasons against:
>
> – not expert in space technology.
>
> – seems to be a management job.
>
> – problem of the programme running out during the next three years and staff reductions envisaged.
>
> – I am just finishing a management job at CERN as DG. Would like to return to technical job.
>
> – My wife wants to stay in Geneva (grand-children, etc.). I would be a bachelor in Paris for 5 years.
>
> – There is a clear link between the next DG ESA and the next DG CERN – perhaps it is me!
>
> – Teillac (the President of Council) has been talking to Schopper. I have the idea that Schopper would not want me around at CERN in any responsible position, e.g. Technical Director or Director of LEP.

Obviously, he was very strongly tempted, because after another meeting in Paris in the following June, he noted:

> 1. If I will accept the job they will look no further. All countries would accept with enthusiasm. They need an answer in October at the latest.
>
> 2. I pointed out two 'problems',
>
>   a) I really wanted to get back to project work (LEP) after my term as DG CERN comes to an end at the end of 1980. Whether this is possible depends on the next DG CERN. Teillac hopes to propose a man in December to CERN Council. The name may be known in October. If so I could talk to him and settle this question.
>
>   b) My wife does not want to leave Geneva and the family. I do not wish to live separately from her during the week. Daily commuting to ESA headquarters is hardly practical from Geneva. Can

headquarters be moved? Answer no – must be in Paris or region according to convention. Could it be at Orly or near airport? In principle yes, but they have spent a lot of money converting a factory into ESA headquarters. This problem seems impossible.

3. Proposals for candidates for DG ESA close in October, after which ESA must decide.
4. New DG could take up post progressively next year.

These extracts show how John used his notebook, not just to record things, but also to try to work out problems by writing down his thoughts. From the last point, John showed that he had considered the possibility that he could move to ESA in April with the new CERN Director-General designate taking over his duties, sharing with Leon, until the end of the year. John and Renie discussed various possibilities, such as having a flat in Paris during the week and returning to Geneva at weekdays, but Renie decided that she did not want to maintain two households and so John reluctantly declined the offer. Unfortunately, the alternatives he noted above were not available to him, since Schopper did not want him to take any further part in LEP, although he did later ask him to join the LEP Machine Advisory Committee, composed mainly of members from other accelerator laboratories, which met occasionally to comment on the design.

Other people were also thinking of jobs for John after ending his spell as Director-General. An editorial in the March 22, 1980 issue of The Economist, headed "Job for the atom-smasher", suggested that John should become the new chairman of the National Nuclear Corporation (NCC), responsible for the building of Britain's nuclear power stations. After describing the mess that this industry was in and dismissing other candidates for various reasons, the article went on "Mr. Adams is – or should be – the man. He is 59, energetic, thinks of himself as the last professional descendant of Isambard Kingdom Brunel; he will certainly be the last boss of a high technology project like CERN to have learnt his engineering on the job, not at a university. ... When there's a good British technological nonsense-basher working abroad, why not at least ask him back?" However, nothing further came of it.

John was also approached to see whether he would be interested in the Mastership of St Catherine's College at Oxford, shortly to become vacant. John went so far so to visit the college, but decided that it was not the sort of life he wanted to lead and so he declined.

At the beginning of 1981 he was asked to join the European Committee for Research and Development and this resulted in his being involved in an assessment of the programme for the four scientific laboratories of the European Commission, known collectively as the Joint Research Centre (JRC). He was appointed to the Governing Board of the JRC 1981 and he noted that the problem child was the ISPRA laboratory in Italy, about which there were criticisms such as "if ISPRA did not exist, nobody would now create it", "there is nothing special for it do in Europe" and "there is a serious lack of motivation, morale is low and discipline is bad". He noted that no clear directive had been given from above and the programme had largely been determined by the laboratory staff. The Board recommended that a proposed experiment in nuclear reactor safety should not be approved, as it would have taken too long and been too expensive for the results expected. John made a proposal that the best solution would be to find a single, unique role for the JRC which would give it unity and purpose. This role had to be clearly definable, important to the Member States and would best be carried out at the European level. Above all, it had to have the wholehearted support of all the Member States. A subject that satisfied these conditions, he suggested, was Safety and the Environment. Therefore he proposed that the JCR should be turned into a European Institute for Safety and Environmental Research, a proposal which was unfortunately not taken up.

In the New Year's Honours of 1981, John was elevated to the rank of Knight Commander of the Order of St. Michael and St. George (KCMG), and so became Sir John. He was invested with the order by the Queen on the 24th March and this time Renie did not miss her celebratory meal, as they were staying at the Royal Society, where he could get out of "the rig" before appearing in the dining room!

Towards the end of 1982, John was informed that he was being proposed to be elected as a foreign member of the Academy of Sciences of the Soviet Union. He had some doubts whether the acceptance of this would be seen as a tacit acceptance of the Soviet system, but, as Wüster put it at John's memorial gathering, "he was firmly convinced that, whatever happened in the political climate, there would never be any justification to sever contacts with those who may be unhappy enough to live in a system which does not give to an individual the same freedom which we enjoy", and so he accepted, being elected early in 1983. In his acceptance letter, after noting some of the eminent scientists who had been similarly honoured, he wrote:

> For me to be included in such a list of names is, I think, a reflection on the way scientific research is carried out these days in such subjects as high energy particle physics ... The discoveries which are now made in this research depend ... not only on theoretical and experimental physicists but also on those who design and build the big accelerators and manage the central laboratories. Although these scientists do not themselves make the discoveries, they nevertheless make major and indispensable contributions to the research.
>
> So when I received the telegram from Professor Alexandrov, the President of the Academy, announcing my election, I therefore felt not only honoured but delighted that the contributions that scientists like me make to the research was being recognised in this way by such a venerable and famous scientific Academy.

This was the same year in which John was asked to join a sub-panel of the High Energy Physics Advisory Panel (HEPAP) concerned with new facilities in the USA, known as the Woods Hole Panel from the small town in Massachusetts where the National Academy of Sciences Study Centre was located. A new facility at Brookhaven, named Isabelle, which was later changed to CBA (Colliding Beams Accelerator), had been approved two years earlier. This involved building a superconducting magnet version of the ISR to collide intense beams of protons with energies up to 400 GeV. However, there had been some difficulties in the development of suitable magnets and the situation at the time of this meeting was that the tunnel and most of the experimental halls had been completed for some time, but no satisfactory production magnets had been produced by the factory set up on the site, although most of the problems seemed to have been solved by then. In surveying the whole field, this panel decided unanimously that the future machine for the USA should be a giant proton collider, with an energy at least 10 TeV per beam, to exceed anything that might be possible to put in the LEP tunnel. This was given the name Superconducting Super Collider (SSC). The panel was divided as to the future of the CBA and the decision was taken, by eleven votes to seven, to cancel it. It is not recorded how John voted, but he had earlier written a note which aimed to show that the completion of the CBA would not have interfered with the construction of such a very big accelerator, because of the staggering of the time scales, and because it might provide useful developments for the bigger project, so one can assume he was amongst the seven. After the decision he wrote:

> The extent of the support which the overall programme gets from the politicians depends on what advantages they will see in it. Not all politicians take the broad national view – they have to watch their votes. The SSC may gain keen national support as a fitting reply to the European challenge. It will also arouse keen competition when it comes to which State it is to be located in. The CBA issue is unlikely to arouse much interest except in the eastern states. However politicians have memories and some may ask why a project so recently declared to be "unique in the world" and "designed to provide a major base for experimentation at the end of this decade" is now to be abandoned in favour of a much bigger machine of the same kind.

The majority decision resulted in the previously unprecedented cancellation of a major accelerator project after many millions of dollars had already been spent on construction work. John's final comment was:

> Admitting a costly mistake is not the best way to get support from another much bigger project. It can easily arouse the feeling that the new project may turn out to be an even bigger mistake.

However, the SSC did survive, though not without delays, and there was keen competition between the States, which was won by Texas putting in a bid to supply money and services that none other could match. Although not the international accelerator envisaged by ICFA earlier, contributions towards its construction are now being sought from other countries, including a large one from Japan.

We have already seen that John was a member of a panel to determine the future of thermonuclear research in Europe, and he had so many other requests that, in answer to a suggestion that he should spend a sabbatical year at Berkeley in the USA, he wrote "unfortunately the news has got around that I am not doing anything and people are asking me to help in various ways in Europe and at CERN." He was also asked to give more lectures than he could fit in.

One of this last commissions was to review the three fields of fast breeder reactors, fusion and high energy physics for the European Commission, in relation to the "Framework" programmes to co-ordinate research in Europe, where he was trying to find methods of streamlining the decision-making process and avoiding the endless references from one committee to another and the necessity to achieve unanimity in all aspects, by increasing the delegated powers for less important points.

So, despite his disappointment at not being able to take a major part in LEP, his abilities were certainly not neglected elsewhere and his activities took him all over the world.

Chapter 14: THE MAN BEHIND THE FAÇADE

John was a difficult man to get to know intimately. While he would talk freely about many subjects, it was almost impossible to get behind the façade he put up to conceal his private life and his emotions. Although I lunched with him several times a week for several years, and we discussed a vast range of subjects, he kept his thoughts on many things to himself. Therefore, to probe further, we have to analyse his character from his reactions on other people.

It has been said that if you want to know something about a man, you should ask his secretary. Susannah Tracy was John's secretary for a large part of the time when he was Executive Director-General at CERN and she told me:

> I was impressed by the breadth of Adams' mind – he had an extraordinary brain. One of the things that struck me first was the way he would listen very carefully to what other people were saying – you know he kept that diary, very religiously, into which he would put various ideas, and you would find, some months later, those ideas coming up, with everything worked out – I think that is what impressed me most when I first came to work for him.
>
> Another thing was his sense of humour, he could always see the funny side of things and that helped to get by in the day work. He was a pleasant person to work for – he would talk to you and always explain to you what he was doing, which doesn't always happen. If there was something I didn't understand or some point I didn't grasp, he was always ready to spend five minutes with me explaining the background – that I appreciated because in a sensitive role like that, secretary to the DG, you need to know what is going on.
>
> A third thing was his sense of organization. One of the first impressions I have of John Adams was when he came in as Project leader for the 300 GeV before it was approved. He would stand in the Council Chamber up at the blackboard and, in his very precise handwriting, he would put up the budget figures for the long-term planning of the SPS and you felt that every delegate would be able to go back to their home country and say "this, gentlemen, is the budget situation and how it will look if we are to build the SPS".
>
> He did face problems, as any DG, but he was always willing to talk to

people, particularly on a one-to-one basis; he was happier if he could solve things without involving too many other people. He would set aside the morning to see people, but in the afternoon he wanted to work undisturbed on the technical side.

He disliked anything impromptu – he always started preparing anything for the official meetings well in advance. As soon as one Finance Committee was finished, he was obviously thinking about the next. Owen Lock was extremely organised, he would always have the agendas prepared for any meeting and then Adams would prepare his speech very carefully.

I got the impression that he knew where he wanted to go and he knew how he was going to do it. Although I wasn't working for him at the time of the 300 GeV, I got the impression that he knew exactly how he was going to steer it through the various committees, ECFA, Committee of Council, Finance Committee and finally the Council and that was a period of about two years. It was very much that Adams was the captain of the ship and he would sail it into port and that would be it.

He was very dependable. If you got him on your side, he would not let anything drift, he would see it through and not let it fall by the wayside. He was very straightforward, very calm, very much attached to his routine in the office; he liked things to be planned and to work – that was his driving force. You didn't bother him with the trivialities of day to day life.

Owen Lock, who was then his Administrative Assistant, confirmed this impression, remarking:

I think I was impressed by the way he organized himself. Every morning, the first person he would see would be his secretary and he would go through the arrangements for the day and sign any correspondence that was there and so on. Then he would see me and we would go through the things I had prepared for him; letters I had drafted or meetings that had to be prepared and by nine o'clock that was it. We were not to disturb him for the rest of the day unless it was very urgent and he was free to concentrate on writing papers and reports or having meetings with his directors or other staff. This was a very efficient way of doing it, both for us and for him as he could settle down for a long time without being interrupted.

The main picture of him in my memory goes back to when I used to go

> home at 5.30 or 6.00 in the winter; he would still be there with his table lamp on, half dark in his office, writing steadily on whatever he was preparing next, His concentration to work steadily was quite impressive. He liked to have everything written down to the last detail. In that respect, it was sometimes a drawback, because in Council or Committee of Council he couldn't always give an impromptu answer to a question and it took him quite a long time to write whatever reports he was preparing.

Other people who worked with him on the more technical side remarked on this meticulous attention to detail. He always wanted to know exactly what was going on. Although he tried to conceal it, it occasionally became obvious that he divided those that worked for him into two categories, those that were for him and those that were neutral or against him. Once he had assured himself of the capabilities of the first type, he trusted them implicitly, but in the second case he would dig down into every detail of their work. An utterance he once made in a state of extreme exasperation has become part of the CERN folklore, that "he built the PS in spite of his staff".

As we have already seen, one of his few failings was this reluctance to hand out verbal compliments and give public recognition to others for the part they had played in his successes, except in a general way, such as "of course this is all due to my loyal team" while implying that, but for his leadership, the outcome would have been quite different. On the other hand, he would often work behind the scenes to try to get recognition for people that he thought deserved it. One example of this was the great number of letters he wrote to Universities in Britain to try and obtain a chair for a member of the staff, who had joined CERN at the beginning, but now wanted to return to the academic life. Also, I doubt whether I would have been awarded my CMG without intensive lobbying of those in power by John.

Another aspect of John was that he would never accept being second best. If he could not dominate a field, he would retire from it. Leon van Hove remarked:

> John always wanted to be the leader and if he couldn't he took a back seat and left it to others. An example was the relations with the local authorities. There used to be annual meetings and in the Lab II days John used to make the speeches. When we became joint Directors-General, I wanted to share the limelight and we started to share them.

After he saw how I could make a speech "off the cuff", while he had to write them out laboriously, he gave up and left them entirely to me.

John also wanted to understand all the technical details, and we have seen how unhappy he was when he thought he would be involved in fast reactors without the nuclear physics background. He later realised that he couldn't be expert in every field as, when building the SPS where computers were going to be more closely involved in the control of the machine than they had been previously, he said to me "I don't understand computers, Michael, so I will have to leave it to you to make a success of it." He was not let down.

John and Renie with two of their grandchildren on holiday.

When John was asked to write an article in a series of "Advice to Young Scientists" in 1970, he pointed out that much of his career had been determined by chance events such as the bombing of Siemens and the deaths of Goward and Bakker, but the important thing was to have a clear sense of direction and to take advantage of all events, adding "If there are not enough events or they are too infrequent, then they can often be stirred up by various means but there is no substitute for a clear sense of direction. An aimless person in a sea of random events takes a very long time indeed to reach any goal." John certainly made sure that he stirred up as many events as possible to achieve his personal goals.

Although John was what Renie called a "workaholic", he did find time for leisure activities. As we have already seen, he loved walking in the mountains in the summer and cross-country skiing in the winter, where his sense of observation enabled him to recognise rare animals and plants that others might have overlooked. He had also been interested in small boats since the Harwell days, when he bought an old air-sea rescue boat which he re-engined to use on the Thames. Later on, when at Culham, he

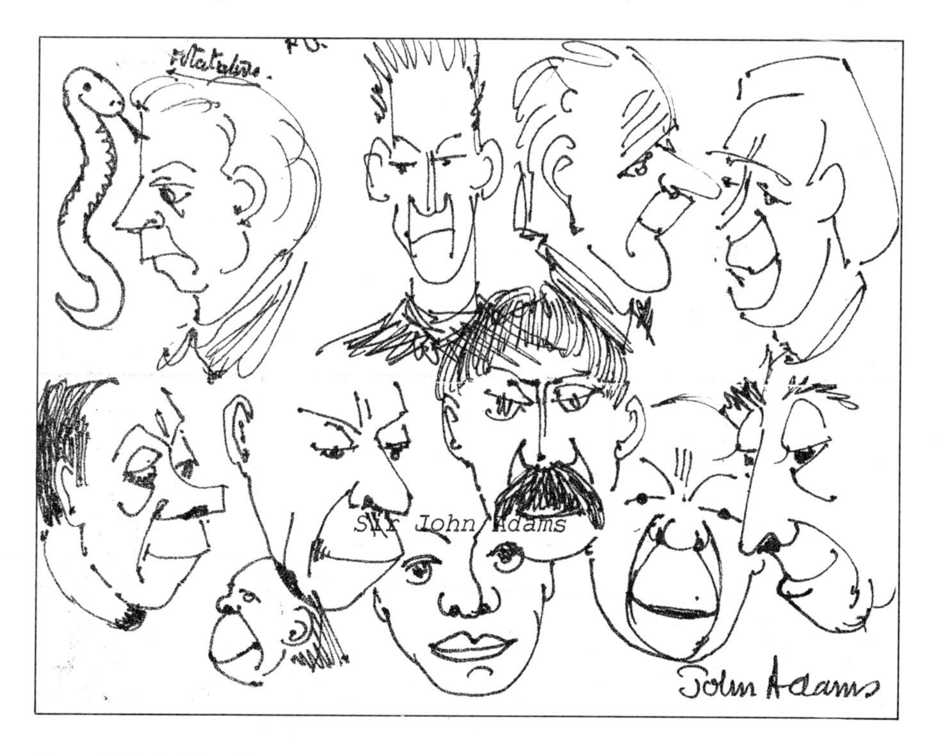

One of John's "doodles".

built a sailing dinghy which they used to trail down to the Costa Brava for holidays. When they returned to Geneva, he bought over the dinghy for use in the south of France, where the Adams family used to borrow a house from a friend for a month each summer. He also liked to work on the various cars he had, doing all the maintenance and some major engine rebuilds, although he always treated them as being mere means of transport rather than being a car enthusiast. Mac Snowden remembers how annoyed John was with himself in the Harwell days when, after overhauling the engine of his Hillman Minx, it wouldn't start because he had got the plug leads on the wrong plugs.

On the more leisurely side, John was an avid reader of books, particularly biographies, history and humour, and liked good music, mainly on the classical side. At the time he was in Abingdon, he went to evening classes to study portrait drawing, an ability he used later during the less interesting parts of official meetings, sketching the participants. There was sometimes a rush to collect the scrap paper under his desk after a meeting, once this pastime became known.

Although he liked good food and wine, John had fairly simple tastes. He often used to complain about the lavish official luncheons he had to attend and when we lunched together his standard fare was a slice of ham and a small salad. When Renie was away, he often dined at our flat in Geneva. If he had business in town, he sometimes arrived early. Despite being invited to have a drink in the lounge, he preferred to sit in the kitchen, where my wife was preparing the dinner, and chat. Her abiding memory of him is sitting there, smoking his pipe and calmly discussing this and that, quite oblivious of the washing that was flapping round his ears, there being nowhere else in the building to dry it.

So far, we have heard a number of opinions about John and his actions but not one from the person closest to him, his wife Renie. She was reluctant to put down her thoughts, as "I knew both sides of his character and I would not like to be entirely frank", but persuaded, she wrote:

> Having studied at school and university in classes with a large proponderence of men I felt that I understood and liked them, but the first time I met John I realized that here was someone different. He was a strikingly good looking man with considerable charm, tall, green eyed and with a shock of curly dark hair. In talking to him one noticed his rather soft voice, ready wit and sense of humour. Some of his remarks were spoken so softly that they were often missed, but when you knew

> him well it was worthwhile listening out for them. Life with John was never boring. We both loved an argument just for the fun of it and if the same argument cropped up again later we often changed sides.
>
> Although he liked chatting people up, particularly pretty women, this was never a cause for worry, for above all he loved his work. In most of the labs in which he worked the "synchrotron widows", as the wives called themselves, became firm friends. They gave up the struggle of competing for their husband's attention with particle accelerators. Luckily these men had the sense to marry interesting women, many of whom gave up their careers to do the job of running the home and bringing up families, thus leaving their men folk free to apply themselves to their chosen work.
>
> John was a good husband and father although he did not have a great deal of time for the children, but he tried hard to keep the weekends free for the family. He gave me my independence, which was terribly important for me. He never interfered if I wanted to teach or help on school boards, or had some other project in hand, and would always give me a help when I needed it. Until he became ill, we had a happy life together and remained a united family.

One of the things that Renie was not entirely frank about was brought up by John Lawson, who told me:

> One thing that used to upset me was the way he treated Renie. She always liked to know what was going on, who was who and the politics involved and so on, and she would come up with some comment and John would put her down in what I thought a very unkind way, particularly when there were visitors there.

Brian Southworth, then editor of the CERN Courier, echoed one of Renie's comments, saying "He would make the most momentous statements in the slow tones of Farmer Giles leaning over the gate and commenting on the state of the weather".

After these observations, let us go back to try to answer the question about John that often comes up: how was it that this man, with the bare minimum of formal education and who started work as an apprentice at the age of sixteen, came to have such an influence in the scientific field?

In following John's career we have been given a glimpse of the way he worked, from the extracts that have been quoted from letters he wrote and from the comments of some of those who came into contact with him.

We have seen how he showed sufficient promise at school for him to have gone on to university, but for his family's financial situation. Instead, he got a good grounding in electrical engineering at the technical college, which enabled him to take advantage of the subsequent opportunities that offered. His experience at TRE was the most valuable in this respect. The people he was working with were mainly new to this field, as he was; they brought academic knowledge and he brought some experience of working in industry. He was accepted as an equal and found that with hard work he could keep up, and in some respects, surpass them, which increased his self-confidence and enabled him to take on greater tasks. His determination to master any subject he was working on and his ability to concentrate on that activity has been remarked upon, also his determination not to be second-best. The second period that shaped his career was at Harwell, where he took on the responsibilities which would have normally only been taken by a much more senior man. Finally, his first years at CERN gave him the experience of managing a multi-national team.

Going back to the first question, one can ask another: How did John influence the scientific field? It was certainly not through his original scientific work, his only publication in this field being the joint letter to the journal Nature on the alternating gradient principle, although later on he wrote many articles on the machines and their uses and on the management of "big science"; that which requires great resources. My explanation is given in the sub-title to this book; that he was an extra-ordinary engineer.

First of all, he was an engineer in the conventional sense, one who designs things, but that was not his greatest strength. He certainly designed things at TRE and in his work on the cyclotron at Harwell, doing the calculations and sketching things out, but after that he was mainly painting the broad picture and assessing and using the detailed design work of others. He was an architect *manqué* and he had an artist's eye as to whether a thing looked right or not. It is interesting to note that his son Christopher inherited this and became an architect himself. I think John's greatest strength, and the clue to much of his success, was his ability to engineer situations, to steer events to follow the path he saw would lead to the result he desired. Susannah's remarks above illustrate this in his dealings with the CERN official committees and I have already described his manner of running technical meetings. Even though he had worked out a situation and come to a conclusion, he would not reveal it all at once, keeping something back which be used to clinch the argument in the future. An example of this was his proposal for changes in the 300 GeV programme. When he first put forward the proposal to build it close by CERN, instead of at another site, he said nothing about using the PS as the injector, instead of the separate linac and booster of the previous design, although there is evidence of him having it in mind from the start. He had certainly written a note giving the implications of this at the same time as his first note on the comparison between "Scheme A" and "Scheme B", but he kept it confidential. Then, after the stormy ECFA meeting criticising the new proposal, he was able to produce this cat out of the bag which, with the reduction in both cost and construction time, won the day. With this endorsement, the Council approved the carrying out of further studies and the approach to the French and Swiss authorities to see whether they would agree to provide the necessary facilities.

Later, when the December Council Meeting had failed to come to a decision and it was postponed until the following February, John wrote to Weisskopf in the USA asking him to visit two of the countries that had reservations, but added:

> But you will need to find pretexts for the visits in order to make the discussions seem accidental. The direct approach may have only an alarming effect and harden attitudes which, from our point of view, already seem rather rigid. It is a delicate matter requiring charm and personal relationships and hence my appeal to you.

Another example of his way of working was recalled by Southworth:

> I saw him operating in a way that I cannot believe was accidental. He would present his ideas not quite fully baked; then his highly intelligent audience concentrated their attention on completing the baking (which this poor engineer had supposedly not quite been able to do) rather than attacking the radically new ideas.

His main lack of success was at the Ministry, where he could not use these techniques so effectively. He tried to work logically in an environment where logic is only a small part of the total picture and, in the short time he persevered, he failed to understand the different time-scales and tactics necessary to get things done.

Let us end on a high note, with some extracts from the eulogies made at the memorial gathering held at CERN on the 27th of April 1984. Anselem Citron recalled the PS days, saying:

> What qualities singled out John Adams as a leader? I shall try and approach this question by commenting on another topic I have heard discussed. Was Adams an optimist or a pessimist? As usual, the answer depends on the meaning you are giving to the terms you use. If an optimist is a person who ignores the difficulties, hoping that with some luck he will get away with it, then Adams was certainly not an optimist. If a pessimist is a person who sees all the difficulties and gets paralysed by them, then Adams was certainly not a pessimist. He would not only see the difficulties, but he would go out of his way to spot them, as we saw in the case of the alternating gradient principle. Then he would set his mind to work to find ways to solve the difficulties, minimise the risks, or build up a second line of defence. After this process, he would go ahead without hesitation and without having to rely on the advice of many people. One of his first acts as director of the PS group was to thank all the consultants and to concentrate on the process of defining the parameters of the PS. I still remember the careful and transparent way in which decisions were reached there under his chairmanship and the atmosphere of calm assurance that emanated from him. This style of John Adams has become, to a large extent, the style of CERN. It has sometimes been criticised, but has made possible the later great achievements about which we shall hear.

Later on, Carlo Rubbia described the proton-antiproton collider work, ending up:

> All these results have been made possible by the marvelous accelerator community that John Adams gathered and forged at CERN. And it is because of this community that our achievements have been made possible. It is often said that the stature of a man, can be measured by the impact he has on the community and by the influence he has on other people in the field. It is certainly correct to say that there would be no modern high energy physics in Europe today without the SPS and the PS, and I truly believe that there would not have been either the PS or the SPS today here without John Adams.

Sir Alec Merrison concluded:

> What else was characteristic of him? First of all the shining integrity of the man: he was an absolutely truthful man in the very widest sense. The second thing was that he was never content with anything less than the very best from himself and that meant that all those that were privileged to be near him and work around him were infected by that spirit, and demonstrated qualities which they had perhaps not realized

in themselves. And then thirdly, something that other people ... have commented upon, the courage of the man. Courage can come from total insensitivity to danger, and that certainly was not John Adams. He was an extremely sensitive man who knew very well the risks he was taking all the time and had the courage to confront those risks. Those are the things which, at least for me, personify him.

Reference

[1] Sir John Adams, 1920-1984: *Speeches made at a memorial meeting held at CERN on 27 April 1984*. CERN, Geneva, 1984.

Chapter 15: FINALE

John had suffered from pains in his chest for some time before he was persuaded to go to see a doctor. Like many people who have been lucky enough to enjoy good health for most of their lives, he had a dislike of going to see a doctor, he mistrusted them and thought that any aches or pains would clear up of their own accord. When he did seek medical advise, towards the end of 1983, lung cancer was diagnosed, and one lung, with a tumour the size of a grapefruit, was removed. Typical of John, he wanted to know all about this subject, and bought books and anatomical diagrams so as to know all that was going on. Things seemed to be progressing well until his heart rate increased and he had difficulty in breathing. He was rushed back to hospital, where it was found that a lymph duct had been severed in the operation and his lung cavity was full of fluid. This was drained, he was put on a diet to reduce the flow and allowed to return home, with the prospect of six weeks of radiotherapy when he was stronger. Unfortunately, he had a relapse and returned to hospital, where he died on the 3rd of March 1984.

John was not outwardly a religious man; he was nominally Church of England and went to church for births, marriages and deaths, but rarely at other times. He hardly ever discussed religion, although Douglas Allen remembers a long discussion he had with him on this subject when they were walking across the Malvern Hills, remarking that John was certainly interested in it.

Although he did not talk much about it, he certainly followed the Christian ethic in his daily life and it must have gone deeper than that because, in the notes John wrote just before his death, he ended up with this statement:

> I have led a remarkably full and happy life. My achievements in the profession I came to follow have been recognised not only in my own country, Britain, but also in the rest of Europe and in many other countries of the world, such as the U.S., U.S.S.R., China and Japan. I have been fortunate to enjoy excellent health up to the present catastrophe and to have been given enormous reserves of energy and enthusiasm. My wife Renie has supported me throughout my hectic life, our children have all married and produced their own families and we remain a remarkably closely knit family. What I am trying to say is that if I die

tomorrow I have absolutely no regrets about the way my life has been spent. I cannot see how I could have done more with it. I cannot now, even if I live on several more years, add very much to what has already been achieved. I am extremely grateful for this life I have been allowed to live. Grateful to whom? Certainly to Renie and all the many colleagues and patrons I have had, but also and above all to the person I have talked to throughout my life, whose advice I have sought at every twist and turn and whose presence and interest I have never doubted. This person is not a human being – he is universally referred to as God.

REQUIESCAT IN PACE

ACKNOWLEDGEMENTS

The writing of this biography would not have been possible without the help I have received from many sources.

First of all, I must thank Renie Adams, who asked me to write it and supplied much vital information. Secondly, Denis Willson, who provided a basic outline from which to start, and Godfrey Stafford, whose biographical note on John for the Royal Society set a standard to be emulated. Thirdly, to Madame R. Rahmy, keeper of the CERN archives, for the help she gave me to access vital documents, together with the press office and photographic staff at CERN for their help.

I would also like to thank all those who talked freely to me about John, or supplied me with written information or corrections to earlier drafts. These include:

Dr. W. D. Allen
The late Prof. E. Amaldi
J. R. Atkinson
R. G. Batt
Rt. Hon. T. Benn
Dr. M. H. Blewett
Dr. F. Bonaudi
Dr. G. Brianti
Prof. W. E. Burcham
C. J. Dadds
Miss P. Dockheer
Lord Flowers
G. H. Hampton
Dr. M. Hine
A. Horseman
Prof. K. Johnsen
Dr. J. Lawson
Dr. Lévy-Mandel
Sir James Lighthill
Dr. O. Lock

P. J. Oliver
Dr. R. S. Pease
Dr. W. H. Penley
The late Lord Penney
Dr. C. Peyrou
The late Dr. T. G. Pickavance
Dr. W. Schnell
M. Snowdon
B. Southworth
Dr. G. Stafford
Miss S. Tracey
The late Prof. L. van Hove
Sir Arthur Vick
Prof. V. Weisskopf
D. Willson
Dr. E. N. J. Wilson
D. Wooton
Dr. K. Zilverschoon

Every attempt has been made to contact the owners of copyright material. However, should any of the material used not be properly acknowledged, the publisher will be glad to provide full acknowledgement in any reprints or future editions of this work.

Appendix 1: THE ACCELERATORS

The first artificial disintegration of the nucleus of the atom was performed by Cockcroft and Walton, using apparatus in which they had accelerated protons to an energy of 800,000 electron-volts by applying a potential difference of 800 kilovolts to two electrodes in an evacuated tube with a source of protons in one electrode. Subsequently direct accelerating machines of this type were developed to provide higher energy particles, spurred by the invention of the electrostatic generator by Van der Graaff, which enabled voltages to several million volts (MV) to be obtained with relatively simple apparatus. However, the practical limitation in terminal voltage for such machines seems to be about 30 MV and to accelerate particles to higher equivalent energies it is necessary to use indirect methods. As early as 1924, Ising proposed to use a succession of electrodes in an evacuated tube to accelerate particles to a high energy by giving them successive kicks by timing the application of voltages between the electrodes, so that the particles obtained an energy several times that they would have obtained from the voltage available. This is the basis of all the subsequent linear accelerators (linacs), but using radio-frequencies to supply the sequential kicks, setting up field in various types of standing-wave and travelling-wave structures. The largest linac built so far, at Stanford University in California, produces electrons with an energy equivalent to being accelerated through 50 GV (50 thousand million volts). However, linear accelerators, except as injectors for the bigger machines, have little part to play in this biography, so we shall concentrate on another way of obtaining high energies.

If a charged particle is subjected to a uniform transverse magnetic field, it will be deflected into a circular orbit, whose radius depends on the energy of the particle and the strength of the magnetic field. If the energy of such a particle is increased, its radius will increase, but so will also its velocity, at least until relativistic effects take place, resulting in it taking exactly the same time to go round the orbit. Ernest Lawrence realised this and built the first machine, which he called a "cyclotron", to use this principle. A cyclotron (Figure 1) consists of an electromagnet with circular poles between which is placed an evacuated chamber containing one or two electrodes, called "dees", from their shape. A voltage, at a fixed radio-frequency (RF), is applied to the dees, setting up an alternating electric field between them, and charged particles, usually protons, are injected into the centre. Those that are injected at the right phase of the RF are accelerated towards a dee by the electric field and perform a semicircle in the magnetic field. With the right choice of RF, the field between the dees will then have reversed, so the particle will be accelerated again. This process will continue, with a gain in energy every semi-circle, so that the path of the particle looks like a spiral, going out to the edge of the magnet, to hit a target or be extracted for external use. However, if the injected particles had any component of velocity in the vertical direction, they would soon get lost by hitting the dees. This is avoided by shaping the poles of the magnet (Figure 2) to provide a vertical restoring force on particles that leave the median plane. This is called "weak-focussing" because the field gradient, n, defined as -dB/B divided by dr/r, must be kept well below unity, otherwise the particles become unstable in the horizontal plane. A cyclotron is limited in the energy of the particles it can produce, to about 30 MeV for protons, since when their velocity is an appreciable fraction of the velocity of light, their mass starts to increase and their rotational frequency slows down, resulting in the particles falling out of synchronism with the RF, so they are not accelerated any further.

This problem was overcome by the invention of the frequency-modu-

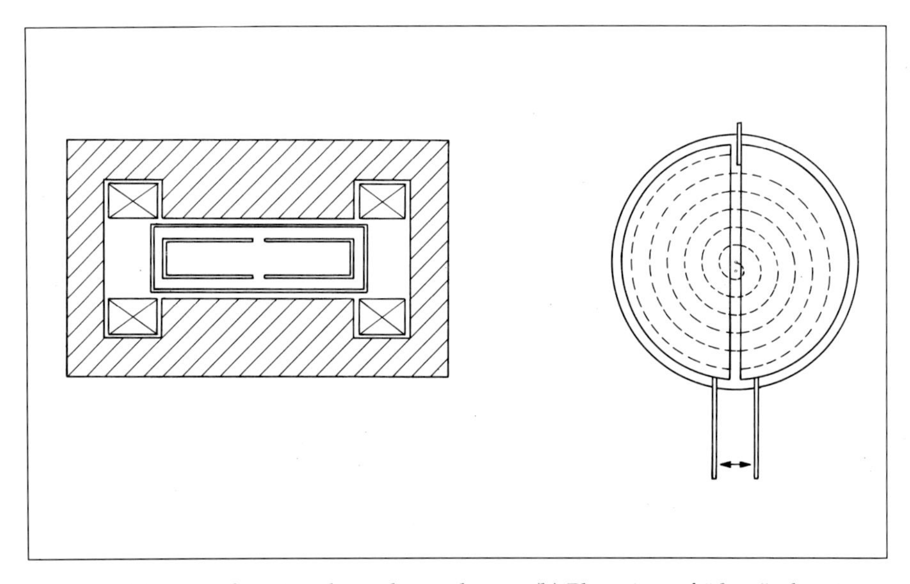

Figure 1 (a) Vertical section through a cyclotron. (b) Plan view of "dees", showing particle orbit as a series of semi-circles of increasing radius.

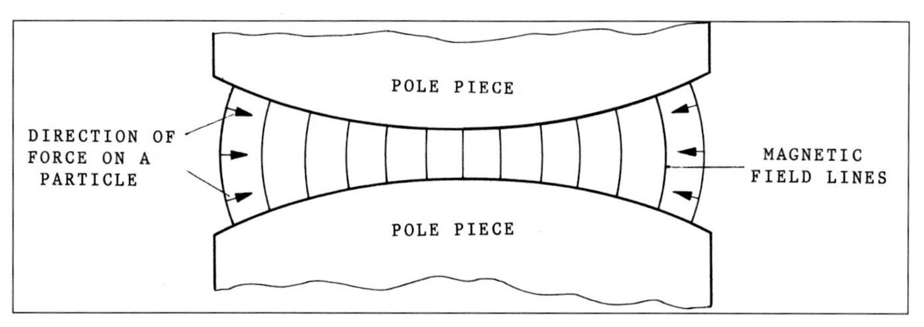

Figure 2 The focussing of particles in the vertical plane by a magnetic field that decreases with increasing radius. Arrows show the direction of the force on a circulating particle.

lated cyclotron or synchro-cyclotron, in which the radio frequency is varied cyclically, starting with the right value for injection and lowering as the particles gain energy, so that they remain in step. Because of this cyclic variation, the particles are accelerated in bursts, rather than continuously, as in the cyclotron. This method of operation was made possible by the discovery of the principle of "phase stability". If the RF voltage is greater than is necessary to give the particles the right energy gain when they pass through the gap between the dees, they will tend to cluster about a phase, known as the stable phase, since a particle at a higher phase will gain too much energy, take longer to perform an orbit and so arrive at a lower phase, and vice versa. Thus the tolerances required on the amplitude and frequency of the RF do not become impossibly tight.

The energy of such a machine is only limited by the size of the magnetic poles and the magnetic field, but as the cost of such a machine tends to go up by a factor greater than the square of the pole diameter, one soon runs into financial difficulties. The Harwell machine had poles 110 inches diameter, and the largest ever built, at Berkeley in USA, had 185 inch poles and could accelerate protons up to 720 MeV.

The next step was the synchrotron, where the magnet, instead of having solid poles, is just the outer annulus. With this, not only the RF but also the magnetic field must be varied cyclically, starting low at injection and rising as the particles gain energy, to keep them on a constant radius. Since such a magnet cannot go down to zero field, due to remanent effects, the particles must be given an initial start in a separate injector accelerator. Since the particles stay around the same path, the massive dees can be replaced by more localised accelerating electrodes. The cost of a synchrotron goes up much less rapidly with energy than a synchro-cyclotron, as the magnet is an annulus of approximately constant cross section, only the circumference going up with energy. However, to reach high beam intensities very large vacuum chambers are needed, because the mutual repulsion between the particles, the space charge, tends to make the beam expand and the focussing forces are only weak, so equilibrium is only obtained with a beam of large cross-section. To allow for this large vacuum chamber (up to 150 cm by 40 cm) the magnet ring had to be correspondingly massive. The largest of this type of accelerator, at Dubna in the USSR, has a magnet system weighing 36,000 tons, to achieve a maximum energy of 10 GeV.

As described in the text, the next stage was the development of strong-focussing. In optics, a convex lens can focus in both planes simultaneously, but the electromagnetic equivalent, a quadrupole, focusses in one

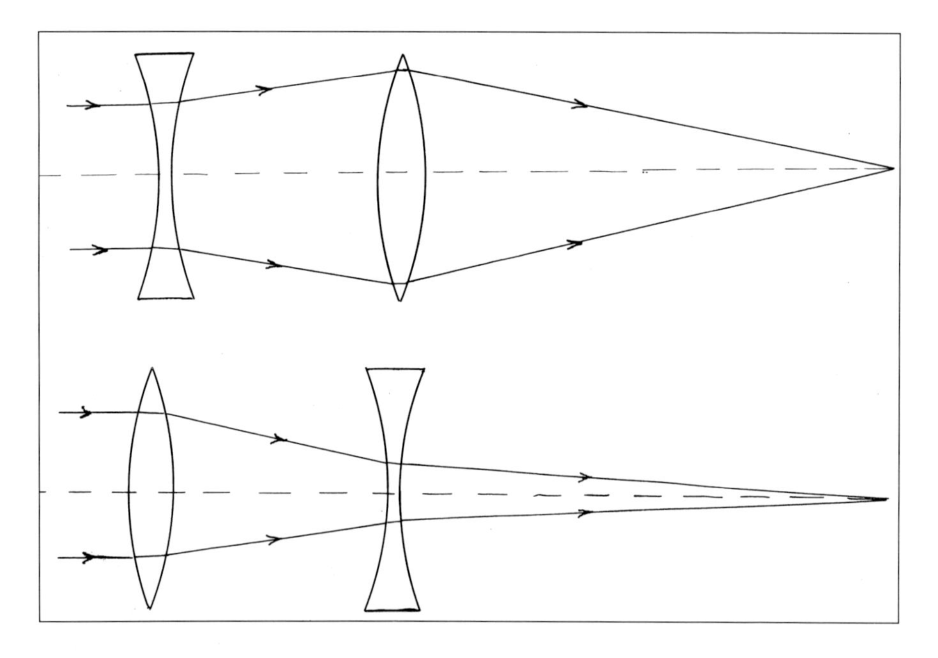

Figure 3 An optical analogue of alternating gradient focussing.

plane and defocusses in the plane at right angles. However, a succession of alternating quadrupoles can be focussing in both planes. This can be understood from the optical analogue shown in Figure 3. The combination of a focussing and a defocussing lens of equal power gives a net focussing effect whichever way they are arranged. As we have seen, this alternating gradient principle enabled the size of the vacuum chamber and the magnets to be reduced dramatically, although not to as great an extent as originally thought. The alternating gradient was obtained in the earlier machines by curving the magnet pole-pieces, as shown in Figure 4, and we have seen how the later use of separate dipole and quadrupole magnets gave greater flexibility, amongst other advantages. Since the beams are now much smaller, the accelerating cavities can be smaller and more efficient.

The next stage was to use colliding beams, to get the advantages discussed in Appendix II. In order to get sufficient interactions, the beams had to built up to much higher intensities by repeated injections of particles into storage rings. One way to get the interactions was to build two interleaved storage rings, as in the case of the ISR, where intense beams of protons circulating in opposite directions in the two rings interacted at eight points. Another way is to collide a beam of particles and anti-particles in a single ring. Since an anti-particle has an opposite electric charge to the particle, it will rotate in the opposite direction in a magnetic field. This was first demonstrated with electrons, since the anti-particle, the position, was relatively easy to produce (see Appendix II), having a rest energy of only

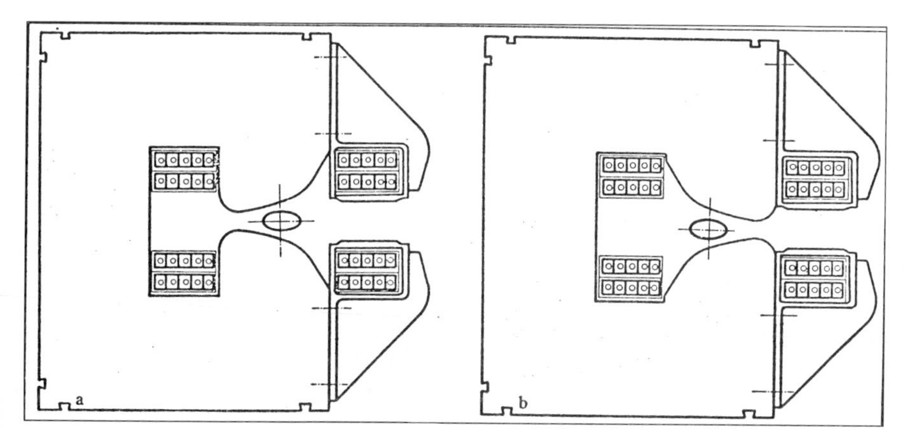

Figure 4 Cross-sections of magnets, showing the pole shapes needed for alternating gradient focussing, using combined function magnets.

half an MeV, so a linac of 100 MeV or so could provide a copious supply. The situation was quite different for the proton, with a rest energy of about 1 GeV. Even using 25 GeV protons, for each million of them hitting a heavy metal target, there would be produced only one or two anti-protons in the forward direction and even these could not be focussed finely enough to be stored in a ring. We have seen that the invention of "stochastic cooling" solved this program, by first collecting the anti-protons in a cooling ring and using feedback systems to reduce the beam size progressively and then "stacking" them until sufficient intensity has been achieved to inject them into the SPS and accelerate them to 270 GeV to collide with the protons which were then injected in the opposite direction round the ring.

Because of its low mass, an electron approaches the speed of light at relatively low energies, which simplifies one problem in electron synchrotons and storage rings, that of varying the radio-frequency during acceleration, but introduces another. When a relativistic particle is made to follow a curved path, it loses energy, emitting photons in the form of "synchrotron radiation". Since the energy loss increases as the fourth power of the particle energy, this becomes serious in an electron-positron collider like LEP at CERN, where even at the initial beam energy of 50 GeV the loss per turn is 200 MeV. The RF system has to make up this loss, so instead of consisting of a few cavities, it has to become sections of a linear accelerator of a size that would have been a major project by itself comparatively few years ago! The increase of energy to about 100 GeV requires the use of superconducting cavities and the doubling of the number of RF sections.

With this synchrotron radiation limitation on circular electron machines, research is going on to try to increase the accelerating field that can be sustained in a linear accelerator, to build machines of reasonable size that could be used to produce beams of electrons of positrons for a "linear collider" of much higher energy than LEP.

The other approach, being used for what will be the world's largest accelerator being built in the USA, the Superconducting Super Collider (SSC), is to use colliding proton beams in a greatly enlarged version of the ISR, but with superconducting magnets to obtain a greater energy for a given size of machine. This will use separate magnets for the two rings. The proposal for a similar machine in the LEP tunnel, called the Large Hadron Collider (LHC), plans to use special magnets with two apertures for the vacuum chambers, so that a single set can accommodate both beams.

Appendix 2: PARTICLE PHYSICS

The Particles

The Greek philosophers had little apparatus to carry out practical experiments (with the notable exception of Archimedes' bath, if the stories are correct), but they had plenty of time to theorise. A number of them, led by Democritus, postulated that while it seemed possible to divide any substance into ever smaller pieces without changing its character, there must be a point where any further subdivision would result in a fundamental change. This smallest representative of the original substance they called an atom. Although there were many different theories as to the nature of matter over the next more than two thousand years, it was not until the Mancunian John Dalton introduced the concept of atomic weight or mass in 1803 that a quantitative basis for the old vague idea of atoms was laid. Many eminent scientists did not accept the atomic theory at first, but were convinced after measurements of the relative masses of different elements had been made in a number of ingenious experiments, leading to the classification of nearly a hundred elements, according to their atomic numbers, in a "Periodic Table" by Mendeléef in 1869.

Meanwhile, the strange "fluid" electricity was attracting scientific attention. Michael Faraday showed that, in electrolysis, a given amount of electricity would transfer masses of elements, according to their atomic numbers, so he suggested that electricity was also "atomic" in the sense of being the flow of numbers of individual particles. This was confirmed in 1897 by Sir J. J. Thompson, who measured the ratio between the electric charge and the mass of this negatively charged particle, the electron, using 'cathode rays' generated by applying a high voltage to electrodes in an evacuated glass tube. Others had found that positively charged particles were also produced in such a tube, with much smaller charge to mass ratios, varying according to the residual gas in the tube, which meant that they were much heavier than the electron. Thompson built what must have been the first mass-spectrometer to carry out more accurate measurements and with this he demonstrated the existence of stable isotopes, elements with the same property but different atomic weights. It had puzzled scientists for some time that the measured relative atomic weight of neon came out as 20.2, while for most other elements the atomic weight were close to whole numbers. Thompson showed that the purified laboratory neon contained two isotopes, with atomic weights 20 and 22, in such proportions to give a mean atomic weight of 20.2

All the work set the scene for the conception of the normally neutral atom being composed of a combination of a relatively heavy positively charged part, neutralised by the required number of light electrons, but it was not clear how were they arranged. The next stage in the unravelling of the secrets of the atom were made possible by a phenomenon known for a long time before, radioactivity. Many substances are naturally radioactive, the best known being radium. Every now and then, an atom in a piece of radium disintegrates, an alpha particle (a helium atom striped of its two electrons) is ejected at high speed. Eventually, after other radioactive decays, only an atom of lead remains. The alpha particles can be detected by the fact that they produce scintillations, tiny flashes of light, when they pass through certain substances, such as zinc sulphide crystals. In 1911, Lord Rutherford showed that if alpha particles went through extremely thin gold foil, most of the particles went almost straight through, but some were violently deflected. From this experiment, Rutherford deduced that the positive part of the atom must be concentrated into a tiny nucleus, less than one thousandth of the size of the atom.

About this time, people were also puzzled by the fact that, for many elements, the positive charge on the nucleus seemed to be about half the atomic weight and the current explanation was that the charge on half the protons were neutralized by electrons in the nucleus, but there were theoretical reasons why this could not be so. In 1920 Rutherford suggested that the nucleus might be made up of a number of positively charged particles combined with a roughly equal number of neutral ones. Practical confirmation of this was given when Chadwick proved the existence of this neutral particle, the neutron, in 1932.

This discovery of the neutron led to a consistent theory. In this, all elements were composed of a nucleus of protons and neutrons, collectively called nucleons, with a cloud of negatively charged electrons circulating round them. The property of the element was determined by the number of protons, and atomic weight by the sum of the protons and neutrons. A number of isotopes of an element were possible, with varying numbers of neutrons. Some of these were stable, but others were unstable and these could decay in a number of ways, after a variable period of time. Some lost a neutron, becoming another isotope of the same element, and in other cases, a neutron emitted an electron, becoming a proton, and so transmuting the element to one of higher atomic number. It was shown that, as a result of the quantum theory, the electrons circulated round the nucleus at discrete energy levels and that to change from one energy level to another an electron had to emit or absorb a photon, the basic unit of electromagnetic energy, and the energy of the photon was a characteristic of the element.

All now seemed clear and in the beginning of the 1930s it was even said that there was nothing further to find and that was the end of nuclear physics. However, there were still a number of thing to be explained. Due to the electromagnetic force, like charges repel each other, so how why do not the protons in a nucleus just fly apart? It was proposed that there must be an attractive force, much stronger than the electromagnetic force, but one that only acts over a very short range. When the protons were very close, this force bound them together, but if they were separated by more than a certain distance, the electromagnetic force would take over and they would fly apart. This 'strong force' also bound the neutrons to the nucleus but, as they had no electric charge, they were not affected by the electromagnetic force. However there were other things that needed explanation. In 1928, P. A. M. Dirac predicted from theoretical grounds that there should be a particle of the mass of the electron but with positive charge. Four years later, C. D. Anderson detected such a particle, named a positron, in an experiment using cosmic rays, the extremely high energy particles that continuously bombard the earth from outer space. Another problem was that, in the decay of certain radioactive materials, the electrons produced were not all of the same energy, as one would expect from the conservation of energy, but covered a whole range of energies, a few with nearly the expected energy, but most with much less. In 1934, Enrico Fermi elaborated an earlier suggestion by Pauli had produced a theory that the missing energy was taken away by a new neutral particle, of zero or negligible mass, which he called the neutrino, "the little neutral one". A particle with no mass and no charge, but only energy, is not only difficult to imagine at first, but also difficult to detect! However this hypothesis solved a number of problems in high-energy physics and so was accepted even though it was many years before neutrinos were observed experimentally in nuclear reactors. Radioactive decay processes are so slow, compared with other nuclear reactions, that they have been given the name of "weak interactions", in contrast to the "strong interactions" when the nuclear binding force is involved. Because neutrinos only interact with atoms through the weak force, they are very penetrating. To explain the reasons for this are beyond this simple story, but one could imagine that the neutrinos, which must travel at the speed of light if they have no mass, pass right through an atom before the slow weak interactions have time to take place. In fact, less than one in a thousand million neutrinos passing through the earth interact with an atom on the way!

Other things turned up to disturb the nice simple theory of the atom. In 1935, Yukawa developed a theory that the strong force could be the result of the exchange of a particle, about 300 times the mass of the electron, between the nucleons and this started a search for such a particle, which he called a mesotron. A particle (later called the muon) of about the right mass was found in some cosmic ray experiments but this did not have the right properties and it was some years before the Yukawa particle, now called the pion, was positively detected. As time went on, and more and more powerful accelerators enabled controlled experiments to be carried out in the energy range only previously available to the unpredictable cosmic rays, many more particles were discovered, most of them existing for only a very short time before decaying into two or more other particles. By 1955, at the time the CERN PS was being built, over thirty of these were listed and the number was still rising. By analogy with the positive electron, a negative proton, the anti-proton, had been propounded and shown to exist, leading to the possibility of anti-matter; if an anti-proton captured a positron it would form an atom of anti-hydrogen, identical chemically to hydrogen, but if this met an atom of normal hydrogen they would annihilate in a flash of energy, far exceeding that from a single fission or fusion!

This proliferation of particles stimulated a search for some unifying principle. They were divided into two main classes: leptons and hadrons. The leptons comprised the rather small family of particles that were not subject to the strong force: the electron, the muon, which behaved like a heavy electron, and the neutrino. The hadrons were further divided into baryons, which included the nucleons, and mesons, such as the pion (the original mesotron). Some symmetries were found between their properties and at that time physicists could be heard muttering "SU(3)" and "the eightfold way". The answer seemed to have been found when Gell-Mann in 1964 showed that it was possible to explain the known hadrons as being built out of a kit of three other particles, and their anti-particles. These he called " quarks", after a phrase from James Joyce's Ulysees "Three quarks for Musther Marks". Baryons would be made up of different combinations of these quarks and mesons from two of them and their anti-particles. However, no free quarks, which would have to have either 1/3 or 2/3 of the charge of an electron, have been detected and theories exist to explain why they cannot exist in isolation.

As always seems to be the case, when a simple explanation has evolved about the basic constituents of matter, complications have to be added to explain the observed facts. To cover these, other quarks had to be thought up and then for it to proposed that each quark should have different forms or "colours". The present list of quarks and leptons is shown in Figure 1. Gell-Mann's original three are now called "up", "down" and "strange". All the stable world can be built up from just the first column of quarks, but the others are needed to explain the transient particles found as a result of nuclear interactions. There is so far no experimental evidence of the existence of the "top" quark but it is expected that it will be found by the next generation of accelerators. It has been possible to get some idea of its mass from recent LEP experiments.

Other efforts were being directed to explain the different "forces" acting on particles, resulting in what is called the "Standard Model". The elec-

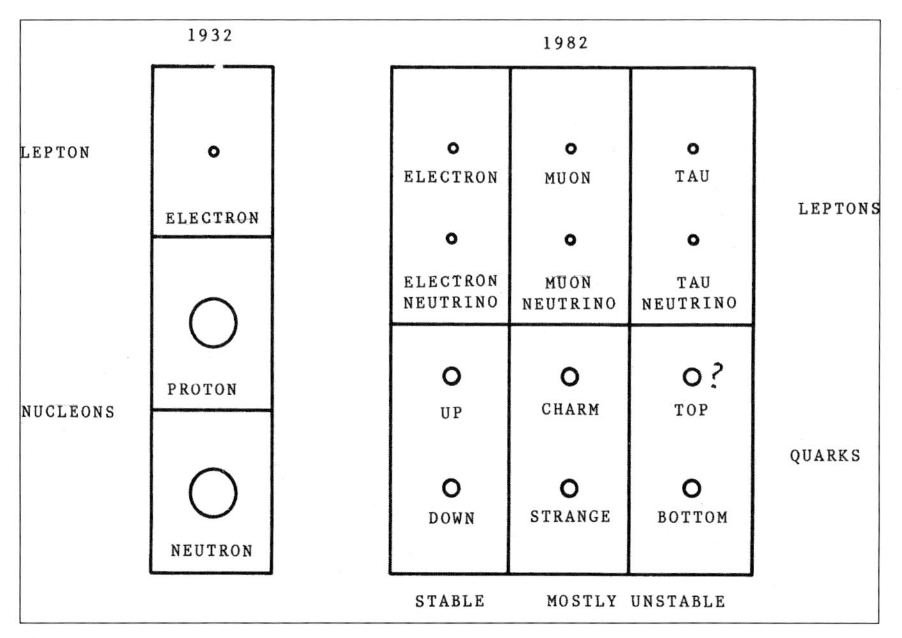

Figure 1 The change in 50 years as to what was considered to be the "elementary" particles of matter.

tromagnetic force is supposed to be the result of the exchange of photons: the force is said to be "mediated" by photons in the jargon. The strong force is mediated by pions in the same way. The Standard Model proposed that the weak force might be mediated by very heavy particles, a neutral one, called the Z-zero and a charged one, the W particle. Using the Einstein relation, $E = mc^2$, the rest mass of a particle is usually expressed as the equivalent energy, which for these particles would be in the range 80 to 100 GeV, and this was one of the determining factors in choosing the energy for LEP, to be able to produce large quantities of them to study their properties. The subsequent proof of the existence of these particles was a boost to the Standard Model, although there remain some inconsistencies, which might be resolved if even higher energy particles, the "Higgs", exist. Even if this can be shown, we have still a long way to go to arrive at the "Great Unified Theory", which would provide the linking to the fourth and weakest force, gravity.

The Experiments

There is no possibility of directly observing atomic particles and so we must trace their paths by the effects they cause. We have already mentioned the scintillation in zinc sulphide crystals in Rutherford's experiments, but better methods were found to give three-dimensional pictures of the passage of particles. The first of these was the photographic emulsion, where the passage of a charged particle causes the ionization of silver atoms along its path, which shows up as a series of dark spots when the emulsion is developed. A more immediate response is given by the cloud chamber, filled with a supersaturated gas. When a charged particle ionizes a gas molecule, it forms a nucleus for the condensation of the gas, leaving a line of tiny drops of liquid in its trail. This is similar to the mechanism that forms the vapour trails following aircraft in a clear sky.

The next development on this line was the bubble chamber, a tank of liquid, usually hydrogen, which is only prevented from boiling by being held at high pressure. If the pressure is suddenly reduced, at the same time as a charged particles pass through the chamber, the liquid will start to boil, but at first only at ionised nuclei left by the passage of the particles. A flash photograph is then taken and the pressure increased to prevent the mass of the liquid from boiling. Most detectors have a superimposed magnetic field, so that the energy of the particles can be determined by their curvature in this field. This can be seen in the bubble chamber photograph and its interpretation shown in Figure 2.

Nowadays, these bubble chambers have been largely replaced by more and more sophisticated detectors, which have been developed over the years, to give greater and greater precision in determining the path of a particle, using electronics to read out the information. These include chambers with tens of thousands of wires strung across them, blocks of thousands of individual pieces of plastic or glass, and special silicon chips. Although they may differ in the way they do it, they nearly all rely on the two basic principles, the ionization or scintillation caused by the passage of a charged particle. As many experiments are looking for rare events in a sea of more common ones, very fast and complicated electronic devices and computers are needed to sort out the desired events from the rest. As

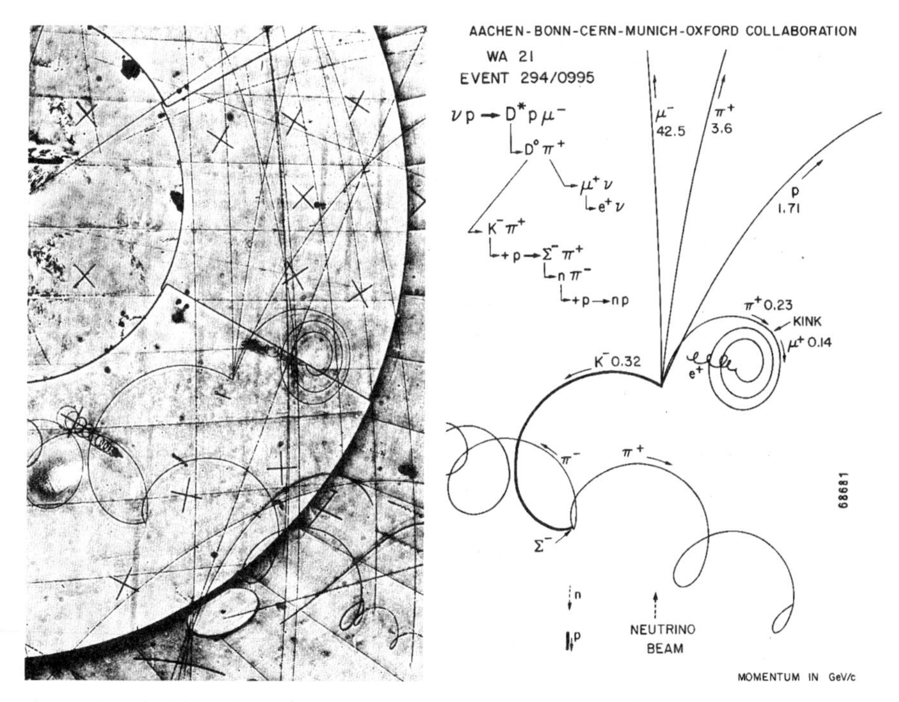

Figure 2 A bubble-chamber picture and its interpretation. The incoming neutrino leaves no trace until it interects with a proton in the liquid, when its energy is given up in generating five charged particles, two of which decay into others while still in the chamber.

the energy of the accelerators increases, so do the size of the detectors needed to carry out the experiments with them. Figure 3 shows the detector for one of the LEP experiments.

Experiments basically observe two types of collisions between particles: elastic and inelastic. The early experiments were examples of elastic collisions. The incoming particle is deflected by the target particle but neither are changed by the "collision". The amount of the deflection gives a measure of the forces involved. In the case of "inelastic" collisions, the states of the incident and target particles change as a result of the collision, possibly producing new particles. To do this, the energy released in the collision must exceed the rest energy (=mass) of the produced particle. This is

Figure 3 Big accelerators need big detectors! The Aleph detector at LEP.

where the colliding beam machines have a great advantage. When a high energy particle strikes a stationary target particle, it knocks the latter forward and an appreciable part of its incident energy is absorbed in doing so. Just as if you throw an apple at a stationary apple, it will knock it away with only slight bruising, but if you throw two apples at each other with the same velocity and they hit head on, they will stop and fall to the ground, with all the energy that they had being used to cause the maximum bruising. Thus particles with equal energies in a colliding beam machine will be much more effective in producing new particles with a high rest energy than a single beam machine using fixed targets.

Power from the Atom

Power can be obtained from two sorts of atomic reactions. One is the fission of heavy elements, where the fission products have less mass than the original atom, the difference being given up as energy. The present atomic power stations use this principle. The other is the fusion of light elements, where the opposite takes place, the combined atom being of less mass than the sum of the fusing elements. This possibility was pointed out in 1919, but it was not until 1934 that Oliphant demonstrated the fusion of two deuterium (Hydrogen 2) nuclei. Later still, in 1942, the fusion of deuterium and tritium (Hydrogen 3) was demonstrated. Subsequent measurements showed that the latter reaction gave the best hope for generating power. However, there is a vast difference between getting a few nuclei to fuse by using accelerators and getting vast numbers of them to do so and give a net output of power. The best approach was to emulate the way fusion occurs in the sun. There, the hydrogen atoms are heated up to such a high temperature that they lose their electrons and the nuclei, in their random motion, have enough energy to overcome the mutual repulsive electromagnetic forces and can come close enough to fuse. This ionized gas is called a plasma. If the required temperature (millions of degrees centigrade) is to be achieved on earth, some method must be found of keeping the plasma from contract with any surface, which would cool it instantly.

Since the nuclei are charged, and so can be bent by a magnetic field, the obvious aim was to form some sort of "magnetic bottle" which would contain the plasma while it was being heated. Very many different configurations of magnetic field that might have worked according to simple theory have been proposed and tried, but the plasma does not behave according to the simple theory and all sorts of instabilities can occur that result in leakage of the plasma or its migration to the walls of the chamber. The most promising configuration seems to be that of a toroidal chamber with crossed magnetic fields to guide and stabilize the plasma, generally known as a "Tokomak" from the original Russian machine of this type. The largest test machine of this type is JET (Joint European Torus) built at the Culham Laboratory by a European consortium.After extensive tests to find the best operating conditions using deuterium alone, it was operated for the first time with tritium added towards the end of 1991. Fusion between the deuterium and tritium nuclei resulted in a net production of power for a few seconds. This successful result means that the tests on JET should provide enough information to enable the next, even larger, machine to be designed to aim for the goal of continuous power production. Even then, there will be a long way to go, and many problems to be solved, before we can plug into fusion power in our homes.

INDEX

Adams, Christopher 25, 74, 94, 168
  Emily 1
  John Albert 1
  Josephine 15, 74, 94
  Katherine 25, 74, 94
  Marjorie 1, 4
Allaby, Jim 113
Allen, Douglas 172
Alvarez, Luiz 56
Amaldi, Edoardo 28, 29, 33, 34, 37, 42, 51, 59, 92, 98, 110
Artzimovich, L. A. 149
Attlee, Rt. Hon. Clement 17
Auger, Pierre 28, 29

Baconnier, Yves 128
Bakker, C. J. 30, 34, 49, 51, 53-56, 59, 60, 62, 70, 164
Bannier, J. H. 91, 118
Benn, Rt. Hon. Antony 81, 84, 86, 89, 93
Beckurts 94
Bernadini, Gilberto 49, 56
Berlin, Isiah 75
Billinge, Roy 118, 125, 143, 144
Binning, Ken 82, 87
Birchall, Vernon 82
Blackett, Lord 79-83
Blewett, Hildred 40, 41, 47-49
  John 40, 41
Bloch, Felix 34, 141
Bohr, Niels 28, 30, 51
Bonaudi, Franco 51, 143
Brianti, Giorgio 44, 49, 118
Brunel, Isambard 146
Bowden, Lord 79
Burgess, Mrs Guy 40
Bulganin 65

Carruthers, Bob 67, 90
Chadwick, Sir James 9, 17, 29, 32
Change, Wen-Yu 150
Churchill, Sir Winston 33
Cherwell, Lord 33, 37, 38
Clarke, Sir Richard 86
Citron, Anselem 169
Cockcroft, J D (Sir John) 17, 18, 20, 22, 24, 25, 29, 32, 25, 37, 40, 45, 53-55, 58, 62, 65, 67, 91
  Lady 40

Cousins, Frank 80, 81

Dahl, Odd 30-32, 34, 39, 41, 42
Dean, Sir Maurice 80, 82, 84, 85
Dee, P. I. 15
de Modzelewska, Eliane 63
de Raad, Baas 117
de Rose, François 59

Fidicaro, Guiseppe 49
Flowers, Lord 101, 113, 114
Fry, Donald 37, 53, 58, 66, 67

Geibel, Hans 47, 48
Gentner, W. 133
Germaine, Pierre 47
Gervaise, Jean 109, 118
Göbel, Klaus 117
Goward, Frank 29, 31, 35, 37-41, 164
Gregory, Bernard 57, 102, 109, 110, 114, 133, 153

Hailsham, Lord 91
Hampton, George 110, 138, 140
Heath, Rt. Hon. Edward 114
Hemmings, Mr & Mrs 5
Hereward, Hugh 48
Hill, Sir John 92
Hine, Mervyn 13, 25, 38-43, 46, 48, 60, 61, 100, 122
Hinton, Sir Christopher 67
Horrisberger, Hans 117
Hughes, J. R. 4

Jentschke, Willi 113, 115, 133, 135, 138, 140, 141
Johnsen, Kjell 31, 39, 40, 42, 43, 46, 55, 61, 99, 100, 104, 108, 117, 119, 143

Keynes, John Maynard 92
Klein, André 119
Kowarski, Lew 30, 43
Kruschev, Nakita 65
Kurchatov, Igor 65
Kurti, Nicolas 74

Lawrence, Ernest 19
Lawson, John 37-39, 41, 65, 167
Levaux, Paul 138

Levy-Mandel, Robert 43, 118, 122, 126, 148
Livingston, Stanley 19
Lock, Owen 152, 162
Lockspeiser, Sir Ben 30, 33, 34
Lovell, Bernard 82
Lloyd Smith 47

Macklusky, Gordon 8
Maddock, Ieuan 82, 84
Marchall, Sir Walter 92
Merrison, Sir Alec 49, 170
Millman, Boris 127, 128

Nikitin, Professor 49

Oliphant, Marc 17

Penncy, Lord 18, 58, 59, 67, 73, 81, 84, 85, 89
Pease, R. S. 70, 71, 73, 76, 90, 92, 149
Peyrou, Charles 56, 57, 60
Picasso, Emilio 147
Pickavance, T. J. 18-21, 58, 113
Peierls, Rudolf 32
Pollock, Jim 2
Powell, C. F. 98

Rabi, Isadore 28
Ramm, Colin 56, 57, 107
Rosset 48
Rowe, A. P. 10
Rubbia, Carlo 142-144, 170

Schmelzer, Christoph 41, 42, 47, 48
Schoch, Arnold 97
Schonland, Basil 58, 59
Schopper, Herwig 143, 146, 147, 155
Schnell, Wolfgang 47, 48
Shaw, Ted 113, 116
Skinner, Herbert 7, 13, 15, 16, 18, 19, 22, 37
Snow, Lord 80, 81
Snowdon, Mac 19, 20, 165
Southworth, Brian 167, 169
Stafford, Godfrey 77, 93, 124
Starr, Arthur 13, 15

Taylor, Brian 76
Teng, Hsaio-ping 150-152
Thatcher, Rt. Hon. Margaret 113-115
Thoneman, Peter 76
Tracy, Susannah 161, 168

van Dardell, Guy 56
Van der Meer, Simon 117, 130, 142-144
van Hove, Leon 98, 136-138, 141, 143, 144, 147, 163
Vick, Sir Arthur 58, 84

Walkinshaw, Bill 37
Warburton, Renie 8
Weisskopf, Victor 60, 62, 63, 98, 141, 169
Wideröe, R. 31
Willson, Denis 70, 73, 74, 89
Wilson, E. N. J. 104, 105, 109, 110, 119
Wilson, Rt. Hon. Harold 79, 80
Wilson, R. R. 100, 105, 118
Wolfson, Lord 75
Wooton, Douglas 8
Wüster, Hans-Otto 93, 118, 128

Zettler, Clement 118